Unthought

Unthought

THE POWER OF THE COGNITIVE NONCONSCIOUS

N. Katherine Hayles

THE UNIVERSITY OF CHICAGO PRESS

CHICAGO AND LONDON

The University of Chicago Press, Chicago 60637
The University of Chicago Press, Ltd., London
© 2017 by The University of Chicago
All rights reserved.

Published 2017
Printed in the United States of America

26 25 24 23 22 21 20 19 18 17   1 2 3 4 5

ISBN-13: 978-0-226-44774-2 (cloth)
ISBN-13: 978-0-226-44788-9 (paper)
ISBN-13: 978-0-226-44791-9 (e-book)
DOI: 10.7208/chicago/9780226447919.001.0001

Library of Congress Cataloging-in-Publication Data

Names: Hayles, Katherine, 1943– author
Title: Unthought : the power of the cognitive nonconscious /
N. Katherine Hayles.
Description: Chicago ; London : The University of Chicago Press,
2017. | Includes bibliographical references and index.
Identifiers: LCCN 2016036767 | ISBN 9780226447742 (cloth : alk.
paper) | ISBN 9780226447889 (pbk. : alk. paper) | ISBN 9780226447919
(e-book)
Subjects: LCSH: Cognition. | Cognition—Social aspects. |
Cognition—Philosophy. | Sociotechnical systems. |
Subconsciousness. | Cognition in literature.
Classification: LCC BF311 .H395 2017 | DDC 154.2—dc23 LC record
available at https://lccn.loc.gov/2016036767

♾ This paper meets the requirements of ANSI/NISO Z39.48–1992
(Permanence of Paper).

For my students

Contents

Acknowledgments

This book could not have come into being had it not been for the generous assistance, intellectual stimulation, and collaborations that helped to catalyze my thoughts (and unthoughts), refine my arguments, and extend my reach. Of primary importance were funding sources that gave me the time to think and write—a sabbatical year from Duke University, a fellowship from the Institute of Advanced Study at the University of Durham, UK, and an appointment as the *Critical Inquiry* Visiting Professor at the University of Chicago. I am pleased to acknowledge their assistance.

During my months in Durham (UK), the windy winter days were warmed by the friends I made there, including Linda Crowe, Mikhail Epstein, Gerhard Lauer, Gerald Moore, Richard Reed, Nicholas Saul, and Veronica Strang. At the University of Chicago, the unparalleled Tom Mitchell was a constant friend, Bill Brown, an inspiring presence, Frances Ferguson, consistently friendly and generous with her time; and the Society of Fellows offered an opportunity for challenging and insightful exchanges. Hank Scotch provided material help and scrupulously accurate copyediting assistance.

Also crucial were scholars who were willing to share their forthcoming work with me before it appeared in print, including Louise Amoore, Ulrik Ekman, Mark Hansen, William Hutchison, Ann-Christina Lange, Luciana Parisi, and Patrick Whitmarsh. Their trust that I would handle these resources appropriately warmed my heart. Providing support and stimulation were my outstanding colleagues at Duke University (Durham, NC), including Rey Chow, Elizabeth Grosz, Mark Hansen, Barbara Herrnstein-Smith, Deborah Jenson, Tim Lenoir, Victoria Szabo, Toril Moi, Antonio Viego, and Robyn Wiegman.

At Duke Kunshan University, where I taught in fall 2015, especially important for my months there was the support of Andrew Field, Haiyan Gao, and Deedra McClearn. Mark Kruse, my co-teacher in our course "Science Fiction/Science Fact," deserves special mention for his remarkable patience, clear expositions, and generosity in providing guidance to me and our students as we probed the complexities of quantum mechanics and relativity theory. Providing help and valuable resources was my research assistant Maryann Murtagh. Marjorie Luesebrink, a lifelong friend, lighted up innumerable conversations and dinners together as we hashed over many of the ideas that found their way into this book.

One of the delights of getting older is to see one's former students flourish in their own right, often with cutting-edge work that exceeds whatever I might have accomplished. I have been fortunate to work with an extraordinarily talented group of young scholars who are now setting the agendas for their fields, including Olivia Banner, Zach Blas, Nathan Brown, Todd Gannon, Amanda Gould, Patrick Jagoda, Melody Jue, Patrick Le Mieux, Kate Marshall, Jessica Pressman, David Rambo, Jenny Rhee, Allen Riddell, David Shepard, John Stadler, and Vidar Thorsteinsson.

I am indebted to many scientists, cultural critics, media theorists, and humanists whose writing and research were crucially important for this project and for my work in general, including Karen Barad, Lauren Berlant, Rosi Braidotti, Jean-Pierre Changeux, Antonio Damasio, Stanislas Dehaene, Gerald Edelman, Danuta Fjellestad, Elizabeth Grosz, Mark Hansen, Donald MacKenzie, Franco Moretti, Luciana Parisi, Garrett Stewart, and Giulio Tononi. Alan Thomas, my editor at the University of Chicago Press, has been a constant friend and colleague now for many years, and I owe him a tremendous debt of gratitude for his steadfast support.

My greatest debt, as always, is to my partner and friend, Nicholas Gessler, who never fails to amaze with his encyclopedic knowledge of everything technical, his dedication to figuring things out for himself, his curiosity about the material world, and his love, warmth, and generosity.

I am grateful to the following for giving me permission to reprint previously published material in this book: most of chapter 2 appeared as "The Cognitive Nonconscious: Enlarging the Mind of the Humanities" in *Critical Inquiry* 4 (4): 783–808 (Summer 2016), and most of chapter 5 also appeared in *Critical Inquiry*, "Cognitive Assemblages: Technical Agency and Human Interactions" 5 (1) (Autumn 2016), © 2016 by The University of Chicago, all rights reserved; portions of chapter 6 appeared as "The Cognitive Nonconscious and Automated Trading Algorithms," in *Parole, écriture, code*, edited by Emmanuele Quinz, translated by Stéphane Vanderhaeghe, Petite Collection ArtsH2H (Paris-Dijon: Les presses du réel, 2015); and portions of chapter 8 appeared as "Cognition Everywhere: The Rise of the Cognitive Nonconscious and the Costs of Consciousness," in *New Literary History* 45.2 (Spring 2014): 199–220.

Transforming How
We See the World

*When he looked at me with his clear, kind, candid eyes, he looked at me
out of a tradition thirteen thousand years old: a way of thought so old,
so well established, so integral and coherent as to give a human being
the unself-consciousness of a wild animal, a great strange creature
who looks straight at you out of his eternal present.*

The epigraph, from Ursula Le Guin's science fiction novel *The Left
Hand of Darkness*, describes the encounter of protagonist Genly Ai with
Faxe, acolyte of the Zen-like cult of the Handdarata and their tradition
of "unlearning" (57). "Given to negatives" (57), the Handdarata would
immediately recognize "unthought" as indicating a kind of thinking
without thinking. There is thought, but before it is unthought: a mode
of interacting with the world enmeshed in the "eternal present" that
forever eludes the belated grasp of consciousness.

"Unthought" may also be taken to refer to recent discoveries in
neuroscience confirming the existence of nonconscious cognitive processes inaccessible to conscious introspection but nevertheless essential for consciousness to function. Understanding the full extent of
their power requires a radical rethinking of cognition from the ground
up. In addition, because the very existence of nonconscious cognitive
processes is largely unknown in the humanities, "unthought" indicates the terra incognita that beckons beyond our received notions of
how consciousness operates. Gesturing toward the rich possibilities
that open when nonconscious cognition is taken into account, "unthought" also names the potent force of conceptualizing interactions
between human and technical systems that enables us to understand

more clearly the political, cultural, and ethical stakes of living in contemporary developed societies.

The first step toward actualizing this potential is terminological
ground clearing about conscious, unconscious, and nonconscious
mental processes.

"Thinking," as used in this book, refers to the thoughts and capabilities associated with higher consciousness such as rationality,
the ability to formulate and manipulate abstract concepts, linguistic
competencies, and so on. Higher consciousness is not, of course, the
whole or indeed even the main part of this story: enhancing and supporting it are the ways in which the embodied subject is embedded
and immersed in environments that function as distributed cognitive
systems. From a cluttered desktop whose complicated topography acts
as an external memory device for its messiness-inclined owner, to the
computer on which I am typing this, to the increasingly dense networks of "smart" technologies that are reconfiguring human lives in
developed societies, human subjects are no longer contained—or even
defined—by the boundaries of their skins.

Part of the book's project is to analyze and explore the nonconscious cognitive assemblages through which these distributed cognitive systems work. In choosing the definite article (*the* cognitive nonconscious), I intend not to reify these systems but rather to indicate
their systemic effects. When my focus is on individual subjects, I will
use the more processually marked term "nonconscious cognitive processes." The power of these assemblages, however, is maximized when
they function as *systems,* with well-defined interfaces and communication circuits between sensors, actuators, processors, storage media,
and distribution networks, and which include human, biological,
technical, and material components. In these instances, I will refer to
the cognitive nonconscious, a term that crucially includes technical
as well as human cognizers. As noted in chapter 5, I prefer "assemblage" over "network" because the configurations in which systems
operate are always in transition, constantly adding and dropping components and rearranging connections. For example, when a person
turns on her cell phone, she becomes part of a nonconscious cognitive
assemblage that includes relay towers and network infrastructures,
including switches, fiber optic cables, and/or wireless routers, as well
as other components. With the cell phone off, the infrastructure is still

in place, but the human subject is no longer a part of that particular cognitive assemblage.

Although nonconscious cognition is not a new concept in cognitive science, neuroscience, and related fields, it has not yet received the attention that I think it deserves. For the humanities, its transformative potential has not yet begun to be grasped, much less explored and discussed. Moreover, even in the sciences, the gap between biological nonconscious cognition and technical nonconscious cognition still yawns as wide as the Grand Canyon on a sunlit morning. One contribution of this study is to propose a definition for cognition that applies to technical systems as well as biological life-forms. At the same time, the definition also excludes material processes such as tsunamis, glaciers, sandstorms, etc. The distinguishing characteristics, as explained in chapter 1, center on interpretation and choice—cognitive activities that both biological life-forms and technical systems enact, but material processes do not. A tsunami, for example, cannot choose to crash against a cliff rather than a crowded beach. The framework I propose, although it recognizes that material processes have awe-inspiring agency, comports neither with vitalism nor panpsychism. Although some respected scholars such as Jane Bennett and Steve Shaviro have given reasons why they find these positions attractive for their purposes, in my view they are not helpful in understanding the specificities of human-technical cognitive assemblages and their power to transform life on the planet.

I see this ongoing transformation as one of the most urgent issues facing us today, with implications that extend into questions about the development of technical autonomous systems and the role that human decision making can and should play in their operation, the environmental devastation resulting from deeply held beliefs that humans are the dominant species on the earth because of their cognitive abilities, and the consequent need for reenvisioning the cognitive capabilities of other life-forms. A correlated development is the spread of computational media into virtually all complex technical systems, along with the pressing need to understand more clearly how their cognitive abilities interact with and interpenetrate human complex systems.

As this framework suggests, another contribution of this study is to formulate the idea of a *planetary cognitive ecology* that includes both human and technical actors and that can appropriately become the focus for ethical inquiry. While traditional ethical inquiries focus on the individual human considered as a subject possessing free will, such perspectives are inadequate to deal with technical devices that operate autonomously, as well as with complex human-technical assemblages in which cognition and decision-making powers are distributed throughout the system. I call the latter cognitive assemblages, and part 2 of this study illustrates how they operate and assesses their implications for our present and future circumstances.

Here is a brief introduction to the book's plan and structure. Part 1 focuses on the concept of nonconscious cognition, with chapter 1 developing a framework for understanding its relation both to consciousness/unconsciousness and material processes. Chapter 2 summarizes the scientific research confirming the existence of nonconscious cognition and locates it in relation to contemporary debates about cognition. Chapter 3 discusses the "new materialisms" and analyzes how these projects can benefit from including nonconscious cognition in their frameworks. As nonconscious cognition is increasingly recognized as a crucial component of human cognitive activity, consciousness has consequently been scrutinized as incurring costs as well as benefits. We can visualize this dynamic as a kind of conceptual seesaw: the higher nonconscious cognition rises in importance and visibility, the lower consciousness declines as the arbiter of human decision making and the dominant human cognitive capability. Chapter 4 illustrates the costs of consciousness through an analysis of two contemporary novels, Tom McCarthy's *Remainder* (2007) and Peter Watts's *Blindsight* (2006).

Part 2 turns to the systemic effects of human-technical cognitive assemblages. Chapter 5 illustrates their dynamics through typical sites ranging from traffic control centers to piloted and autonomous drones. Chapter 6 focuses on autonomous trading algorithms, showing how they require and instantiate technical autonomy because the speeds at which they operate far transcend the temporal regimes of human decision making. This chapter also discusses the implications of these kinds of cognitive assemblages, particularly their systemic effects on destabilizing the global economy. Chapter 7 explores the ethical implications of cognitive assemblages through a close reading of Colson Whitehead's novel *The Intuitionist*. Chapter 8 expounds on the utopian potential of cognitive assemblages and extends the argument to the digital humanities, proposing that they too may be considered as cognitive assemblages and showing how the proposed framework of nonconscious cognition affects how the digital humanities are understood and evaluated.

In conclusion, I want to present a few takeaway ideas that I hope every reader of this book will grasp: most human cognition happens outside of consciousness/unconsciousness; cognition extends through the entire biological spectrum, including animals and plants; technical devices cognize, and in doing so profoundly influence human complex systems; we live in an era when the planetary cognitive ecology is undergoing rapid transformation, urgently requiring us to rethink cognition and reenvision its consequences on a global scale. My hope is that these ideas, which some readers may regard as controversial in part or whole, will nevertheless help to initiate conversations about cognition and its importance for understanding our contemporary situations and moving us toward more sustainable, enduring, and flourishing environments for all living beings and nonhuman others.

PART 1

THE COGNITIVE NONCONSCIOUS AND THE COSTS OF CONSCIOUSNESS

CHAPTER 1

Nonconscious Cognitions: Humans and Others

Rooted in anthropocentric projection, the perception that consciousness and advanced thinking necessarily go together has centuries, if not millennia, of tradition behind it. Recently, however, a broad-based reassessment of the limitations of consciousness has led to a correspondingly broad revision of the functions performed by other cognitive capacities and the critical roles they play in human neurological processes. Consciousness occupies a central position in our thinking not because it is the whole of cognition but because it creates the (sometimes fictitious) narratives that make sense of our lives and support basic assumptions about worldly coherence. Cognition, by contrast, is a much broader capacity that extends far beyond consciousness into other neurological brain processes; it is also pervasive in other life forms and complex technical systems. Although the cognitive capacity that exists beyond consciousness goes by various names, I call it nonconscious cognition.

Perhaps no areas are more rife with terminological disparities than those dealing with consciousness; rather than sort through centuries of confusions, I will try to make clear how I am using the terms and attempt to do so consistently throughout. "Consciousness," as I use the term, comprises core or primary consciousness (Damasio 2000; Dehaene 2014; Edelman and Tononi 2000), an awareness of self and others shared by humans, many mammals, and some aquatic species such as octopi. In addition, humans and (perhaps) a few primates manifest extended (Damasio 2000) or secondary (Edelman and Tononi 2000) consciousness, associated with symbolic reasoning, abstract thought, verbal language, mathematics, and so forth (Eagleman 2012; Dehaene 2014). Higher consciousness is associated with the autobiographical self (Damasio 2012, 203–07), reinforced through the verbal monologue that plays in our heads as we go about our daily business; that monologue, in turn, is associated with the emergence of a self aware of itself as a self (Nelson, in Fireman, McVay, and Flanagan 2003, 17–36). Recognizing that the cognitive nonconscious (in his terms, the protoself) can create a kind of sensory or nonverbal narrative, Damasio explains how the narratives become more specific when melded with verbal content in higher consciousness. "In brains endowed with abundant memory, language, and reasoning, narratives . . . are enriched and allowed to display even more knowledge, thus producing a well-defined protagonist, the autobiographical self" (Damasio 2012, 204). Whenever verbal narratives are evoked or represented, this is the mental faculty that makes sense of them.[1]

Core consciousness is not sharply distinguished from the so-called "new" unconscious (in my view, not an especially felicitous phrase), a broad environmental scanning that operates below conscious attention (Hassin, Uleman, and Bargh 2005). Suppose, for example, you are driving while thinking about a problem. Suddenly the car in front brakes, and your attention snaps back to the road. The easy and continuous communication between consciousness and the "new" unconscious suggests that they can be grouped together as modes of awareness.[2]

In contrast, nonconscious cognition operates at a level of neuronal processing inaccessible to the modes of awareness but nevertheless performing functions essential to consciousness. The last couple of decades in neuroscientific research show that these include integrating somatic markers into coherent body representations (Damasio 2000), synthesizing sensory inputs so they appear consistent across time and space (Eagleman 2012), processing information much faster than can consciousness (Dehaene 2014), recognizing patterns too complex and subtle for consciousness to discern (Kouider and Dehaene 2007), and drawing inferences that influence behavior and help to determine priorities (Lewicki, Hill, and Czyzewska 1992). Perhaps its most important function is to keep consciousness, with its slow uptake and limited processing ability, from being overwhelmed with the floods of interior and exterior information streaming into the brain every millisecond.

The point of emphasizing nonconscious cognition is not to ignore the achievements of conscious thought, often seen as the defining characteristic of humans, but rather to arrive at a more balanced and accurate view of human cognitive ecology that opens it to comparisons with other biological cognizers on the one hand and on the other to the cognitive capabilities of technical systems. Once we overcome the (mis)perception that humans are the only important or relevant cognizers on the planet, a wealth of new questions, issues, and ethical considerations come into view. To address these, this chapter offers a theoretical framework that integrates consciousness, nonconscious cognition, and material processes into a perspective that enables us to think about the relationships that enmesh biological and technical cognition together.

Although technical cognition is often compared with the operations of consciousness (a view I do not share, as discussed below), the processes performed by human nonconscious cognition form a much closer analogue. Like human nonconscious cognition, technical cognition processes information faster than consciousness, discerns patterns and draws inferences and, for state-aware systems, processes inputs from subsystems that give information on the system's condition and functioning. Moreover, technical cognitions are designed specifically to keep human consciousness from being overwhelmed by massive informational streams so large, complex, and multifaceted that they could never be processed by human brains. These parallels are not accidental. Their emergence represents the exteriorization of cognitive abilities, once resident only in biological organisms, into the world, where they are rapidly transforming the ways in which human cultures interact with broader planetary ecologies. Indeed, biological and technical cognitions are now so deeply entwined that it is more accurate to say they interpenetrate one another.

The title of part 1, the cognitive nonconscious, is meant to gesture toward the systematicity of human-technical interactions. In part 2, I will refer to these as cognitive assemblages. *Assemblage* here should not be understood as merely an amorphous blob. Although open to chance events in some respects, interactions within cognitive assemblages are precisely structured by the sensors, perceptors, actuators, and cognitive processes of the interactors. Because these processes can, on both individual and collective levels, have emergent effects, I will use *nonconscious cognition(s)* to refer to them when the emphasis is on their abilities for fluid mutations and transformations. The more reified formulation indicated by the definite article (*the* cognitive nonconscious) is used when the systematicity of the assemblage is important. I adopt this form for my overall project because the larger implications of cognitive assemblages occur at the systemic rather than individual levels. As a whole, my project aims to chart the transformative perspectives that emerge when nonconscious cognitions are taken fully into account as essential to human experience, biological life, and technical systems.

Although my focus is on biological and technical cognitions that function without conscious awareness, it may be helpful to clarify my position relative to the cognitivist paradigm that sees consciousness operating through formal symbol manipulations, a framework equating the operations of human minds with computers. Clearly humans can abstract from specific situations into formal representations; virtually all of mathematics depends on these operations. I doubt, however, that formal symbol manipulations are generally characteristic of conscious thought. Jean-Pierre Dupuy (2009), in his study arguing that cognitive science developed from cybernetics but crucially transformed its assumptions, characterizes the cognitivist paradigm not as the humanization of the machine (as Norbert Weiner at times wanted to position cybernetics) but as the mechanization of mind: "The computation of the cognitivists . . . is symbolic computation. The semantic objects with which it deals are therefore all at hand: they are the mental representations that are supposed to correspond to those beliefs, desires, and so forth, by means of which we interpret the acts of ourselves and others. Thinking amounts, then, to performing computations on these representations" (Dupuy 2009, 13).

As Dupuy shows, this construction is open to multiple objections. Although cognitivism has been the dominant paradigm within cognitive science throughout the 1990s and into the twenty-first century, it is increasingly coming under pressure to marshal experimental evidence showing that brains actually do perform such computational processes in everyday thought. So far, the results remain scanty, whereas experimental confirmation continues to grow for what Lawrence Barsalou (2008) calls "grounded cognition," cognition supported by and entwined with mental simulations of modal perceptions, including muscle movements, visual stimuli, and acoustic perceptions. In part this is because of the discovery of mirror neuron circuits in human and primate brains (Ramachandran 2012), which, as Miguel Nicolelis (2012) has shown in his work on Brain-Machine-Interfaces (BMI), play crucial roles in enabling humans, primates, and other animals to extrapolate beyond bodily functions such as limb movements into prosthetic extensions.

One aspect of these controversies is whether neuronal processes can in themselves be understood as fundamentally computational. Dissenting from the computationalist view, Walter J. Freeman and Rafael Núñez argue that "action potentials are not binary digits, and neurons do not perform Boolean algebra" (1999, xvi). Eleanor Rosch, in "Reclaiming Concepts" (Núñez and Freeman 1999, 61–78) carefully contrasts the cognitivist paradigm with the embodied/embedded view, arguing that empirical evidence is strongly in favor of the latter. Amodal symbolic manipulation, as Barsalou (2008) characterizes the cognitivist paradigm, depends solely on logical formulations unsupported by the body's rich repertoire of physical actions in the world. As numerous researchers and theorists have shown (Lakoff and Johnson 2003; Dreyfus 1972, 1992; Clark 2008), embodied and embedded actions are crucial in the formation of verbal schema and intellectual comprehension that express themselves through metaphors and abstractions, extending out from the body to sophisticated thoughts about how the world works.

My comparison between nonconscious cognition in biological life-forms and computational media is not meant to suggest, then, that the processes they enact are identical or even largely similar, because those processes take place in very different material and physical contexts. Rather, they perform similar *functions* within complex human and technical systems. Although functionalism has sometimes been used to imply that the actual physical processes do not matter, as long as the results are the same (for example, in behaviorism and some versions of cybernetics), the framework advanced here makes context crucial to nonconscious cognition, including the biological and technical milieu within which cognitions take place. Notwithstanding the profound differences in contexts, nonconscious cognitions in biological organisms and technical systems share certain *structural* and *functional* similarities, specifically in building up layers of interactions from low-level choices, and consequently very simple cognitions, to higher cognitions and interpretations.

Exploring these structural parallels requires a good deal of ground clearing to dispense with lingering questions such as whether machines can think, what distinguishes cognition from consciousness and thought, and how cognition interacts with and differs from material processes. Following from these fundamental questions are further issues regarding the nature of agencies that computational and biological media possess, especially compared with material processes, and the ethical implications when technical cognitive systems act as autonomous actors in cognitive assemblages. What criteria for ethical responsibility are appropriate, for example, when lethal force is executed by a drone or robot warrior acting autonomously? Should it focus on the technical device, the human(s) who set it in motion, or the manufacturer? What perspectives offer frameworks robust enough to accommodate the exponentially expanding systems of technical cognitions and yet nuanced enough to capture their complex interactions with human cultural and social systems?

Asking such questions is like pulling a thread dangling from the bottom of a sweater; the more one pulls, the more the whole fabric of thinking about the significance of biological and computational media begins to unravel. Parts 1 and 2 pull as hard as they can on that thread and try to reweave it into different patterns that reassess the nature of human and technical agencies, realign human and technical cognitions, and investigate how these patterns present new opportunities and challenges for the humanities.

THINKING AND COGNITION

The first twist in knitting these new patterns is to distinguish between thinking and cognition. Thinking, as I use the term, refers to high-level mental operations such as reasoning abstractly, creating and using verbal languages, constructing mathematical theorems, composing music, and the like, operations associated with higher consciousness. Although Homo sapiens may not be unique in these abilities, humans possess them in greater degree and with more extensive development than other species. Cognition, by contrast, is a much broader faculty present to some degree in all biological life-forms and many technical systems. This vision overlaps with the position that Humberto Maturana and Francisco Varela articulated in their classic work on cognition and autopoiesis (1980). It also aligns with the emerging science of cognitive biology, which views all organisms as engaging in systematic acts of cognition as they interact with their environments. The field, named by Brian C. Goodwin (1977), has subsequently been developed by the Slovakian scientist Ladislav Kováč (2000, hereafter referred to as "FP"; 2007), who has been instrumental in codifying its principles and exploring its implications.

Cognition as formulated in cognitive biology employs some of the same terms as mainstream views but radically alters their import. Traditionally, cognition is associated with human thought; William James, for example, noted that "cognition is a function of consciousness" ([1909] 1975, 13). Moreover, it is often defined as an "act of knowing" that includes "perception and judgment" ("Cognition," in *Encyclopedia Britannica*, www.britannica.com/topic/cognition-thought -process). A very different perspective informs the principles of cognitive biology. Consider, for example, Kováč's observation that even a unicellular organism "must have a certain minimal knowledge of the relevant features of the environment," resulting in a correspondence, "however coarse-grained and abstract," between these features and the molecules of which it is comprised. He concludes, "In general, at all levels of life, not just at the level of nucleic acid molecules, a complexity, which serves a specific function . . . corresponds to an *embodied knowledge*, translated into the constructions of a system. The environment is a rich set of potential niches: each niche is a problem to be solved, to survive in the niche means to solve the problem, and the solution is the embodied knowledge, an algorithm of how to act in order to survive" ("FP," 59). In this view cognition is not limited to humans or organisms with consciousness; it extends to all life-forms, including those lacking central nervous systems, such as plants and microorganisms.

The advantages of this perspective include breaking out of an anthropocentric view of cognition and building bridges across different phyla to construct a comparative view of cognition. As formulated by Pamela Lyon and Jonathan Opie (2007), cognitive biology offers a framework consistent with empirical results: "Mounting evidence suggests that even bacteria grapple with problems long familiar to cognitive scientists, including: integrating information from multiple sensory channels to marshal an effective response to fluctuating conditions; making decisions under conditions of uncertainty; communicating with conspecifics and others (honestly and deceptively); and coordinating collective behavior to increase the chances of survival."[3] Kováč calls the engagement of a life-form with its environment its *onticity,* its ability to survive and endure in changing circumstances. He observes that "life incessantly, at all levels, by millions of species, is 'testing' all the possibilities of how to advance ahead" ("FP," 58). In a playful extension of this reasoning, he imagines a bacterial philosopher confronting the same issues concerning its onticity as a human, asking whether the world exists, and if so, why there is something rather than nothing. Like the human, the bacterium can find no absolute answers within its purview; it nevertheless pursues "its onticity in the world" and accordingly "is already a *subject,* facing the world as an object. At all levels, from the simplest to the most complex, the overall construction of the subject, the embodiment of the achieved knowledge, represents its *epistemic complexity*" ("FP," 59). The sum total of the world's epistemic complexity is continually increasing, according to Kováč, advanced by the testing of what he calls the beliefs of organisms: "only some of the constructions of organisms are embodied knowledge, the others are but *embodied beliefs. . . .* If we take a mutation in a bacterium as a new belief about the environment, we can say that the mutant would sacrifice its life to prove its fidelity to that belief" ("FP," 63). If it continues to survive, that belief becomes converted into embodied knowledge and, as such, is passed along to the next generation.

Comparing traditional and cognitive biology perspectives shows that the same words attain very different meanings. *Knowledge,* in the traditional view, remains almost entirely within the purview of awareness and certainly within the brain. In cognitive biology, on the contrary, it is acquired through interactions with the environment and embodied in the organism's structures and repertoire of behaviors. *Belief* in the traditional view is a position held by a conscious being as a result of experience, ideology, social conditioning, and other factors. In the cognitive biology view, it is a predisposition toward the environment that has not yet been confirmed through ongoing interactions testing its robustness as an evolutionary response to fluctuating conditions. Finally, *subject* in the traditional view is taken to refer to humans or at least conscious beings, while in the cognitive biology view it encompasses all life forms, even humble unicellular organisms.

PLANT SIGNALING AND CLAIMS FOR PLANT INTELLIGENCE

A convenient site to explore the complex interactions that arise when these perspectives on cognition confront traditional views of intelligence is the world of plants. In a recent *New Yorker* article, Michael Pollan summarizes research that explores homologies between "neurobiology and phytobiology," specifically that plants are "capable of cognition, communication, information processing, computation, learning and memory" (Pollan 2013, 1). The claims are made explicit in a 2006 article in *Trends in Plant Science* (Brenner et al.). Positioned as a review article, the piece is also a polemical manifesto aiming to establish the field of plant neurobiology, arguing that many of the complexities of plant signaling strongly parallel animal neurobiology. As the authors recognize, plant "intelligence" had become a lightning rod for controversy since the 1973 pop science book *The Secret Life of Plants* by Peter Tompkins and Christopher Bird, which made extraordinary claims with little evidence. As a result, many plant scientists wanted to distance themselves as much as possible from claims about plant "intelligence," including the assertion that plants are somehow attuned to human emotional states. Brenner et al. suggest that as a result, many plant biologists refused even to consider parallels between plant responses and animal neurology, practicing "a form of self-censorship in thought, discussion and research that inhibited asking relevant questions" (415).

However justified this comment, the Brenner article itself manifests rhetorical and argumentative strategies that exhibit a deep ambivalence. On the one hand, the authors want to document research showing how complex and nuanced are the mechanisms that underlie individual and communal plant behaviors; on the other, they inadvertently reinstall the privilege of animal intelligence by implying that the more plant signaling resembles animal neurobiology, the stronger the case that it is *really* intelligence. The ambivalence is apparent in the sidebar tracing the etymology of the term "neuron" back to Plato and the Greeks, where "'neuron' means "anything of a fibrous nature" (414). By this definition, plants clearly do have neurons, but in the usual sense of the term (cells with nuclei and axons that communicate using neurotransmitters), they do not. A similar ambivalence is apparent in how they define intelligence; by insisting on the word, they create a rhetorical tension between what they seem to be claiming and what they are actually saying. Offering first a definition of plant intelligence (from Trewavas 2005) as "'adaptively variable growth over the lifetime of a plant'" (414), they expand on it, adding an emphasis on processing information and making decisions: "an intrinsic ability to process information from both abiotic and biotic stimuli that allows optimal decisions about future activities in a given environment" (414).

In my view, this definition offers important clues for reenvisioning cognition (a trajectory I was already following before reading the Brenner article), as well as providing a case study in why it is better to avoid using "intelligence" for nonhuman (and technical) cognitions. As Pollan documents, "many plant scientists have pushed back hard" against what they (mis)understood to be the argument. He notes that thirty-six plant biologists issued a rebuttal to the Brenner piece, also published in *Trends in Plant Science*. The refutation opens with this salvo: "We begin by stating simply that there is no evidence for structures such as neurons, synapses or a brain in plants" (qtd in Pollan, 3). Pollan points out that "no such claim had actually been made—the manifesto had spoken only of 'homologous' structures—but the use of the word 'neurobiology' in the absence of actual neurons was apparently more than the scientists could bear" (3). This rather snide comment (revealing Pollan's own sympathies) does not, in my view, do justice to the complexities of the situation. The issue is not what plant scientists can bear, but how traditional views of intelligence interact with and complicate research that challenges (and perhaps also inadvertently reinstalls) the anthropocentric perspective of what intelligence is. Daniel Chamovitz, for example, while insisting on the remarkable abilities of plants to sense and respond to their environments, argues that "the question . . . should not be whether or not plants are *intelligent*—it will be ages between we all agree on what that term means; the question should be, 'Are plants aware?' and, in fact, they are" (2013, 170). Indeed, Pollan himself points out that "the controversy is less about the remarkable discoveries of recent plant science than about how to interpret and name them; whether behaviors observed in plants which look very much like learning, memory, decision-making and intelligence deserve to be called by those names or whether those words should be reserved exclusively for creatures with brains" (4).

For an analogy, I think of Gillian Beer's brilliant study in *Darwin's Plots: Evolutionary Narrative in Darwin, George Eliot, and Nineteenth-Century Fiction* (1983) tracing the struggle in Darwin's *The Origin of Species* between his view of evolution as a process with no foreordained end and the teleological worldview embedded in the Christian-oriented language he inherited and upon which he instinctively drew. Through a series of close readings, Beer traces in Darwin's metaphors, sentence structures, and rhetorical strategies his desire to articulate a new vision through language saturated with the old. A similar struggle informs the Brenner article; although it is true that the scientists who objected to the article's claims did misread it in a literal sense, they were reacting to the kind of ambivalence noted above between actual evidence and insinuations carried through such tactics as redefining "neuron." In this sense, they accurately discerned the article's double intent to draw upon the cachet of "intelligence" as an anthropocentric value while simultaneously revising the criteria for what constitutes intelligence.

Since plants make up 99 per cent of the planet's biomass, the issue is not trivial across a range of sites, including the question Christopher D. Stone ([1972] 2010) posed decades ago of whether trees should have legal standing. My own clear preference is to create a framework that is both robust and inclusive, and I see no way to exclude plants without sacrificing conceptual coherence (not to mention ignoring the wealth of evidence documenting their remarkable abilities to respond to changing environments).[4]

Nevertheless, assuming that one wanted to draw the line separating cognitive organisms from the noncognitive differently, most aspects crucial to my argument could still be included: the reevaluation of cognition as distinct from consciousness; the recognition that cognitive technologies are now a potent force in our planetary cognitive ecology; and the rapidly escalating complexities created by the interpenetration of cognitive technologies with human systems. These, in my view, are not debatable, while the arguments about plants occupy a less central (although still important) role in my own priorities. I recognize, then, that locating the boundary between the cognitive and noncognitive may be contested, and that different perspectives will lead to conclusions other than those that I endorse. The crucial point for me is less where the line is drawn than that the core issues mentioned above are recognized as critical to our contemporary situation. For me, another important point is the role that humanistic inquiry can play in this arena. Because reenvisioning cognition occurs along a broad interdisciplinary front fraught with linguistic as well as conceptual complexities, the humanities, with their nuanced understanding of rhetoric, argument, and interpretation, are well positioned to contribute to the debate.

I conclude this section with a brief acknowledgement of how complex plant cognition is, where "cognition" here refers to the ways plants sense information from their surroundings, communicate within themselves and to other biota, and respond flexibly and adaptively to their changing environments. Their "'sessile life style'" (Pollan, 4–5—"sessile" refers to organisms attached directly to a substrate, for example, corals and almost all plants) includes more than a dozen senses, among them kin recognition, detection of chemical signals from other plants, and analogues to the five human senses. Pollan explains how kin recognition has been observed to work: "Roots can tell whether nearby roots are self or other, and if other, kin or stranger. Normally, plants compete for root space with strangers, but, when researchers put closely related Great Lakes sea-rocket plants (*cakile edentual*) in the same pot, the plants restrained their usual competitive behaviors and shared resources" (Pollan, 5). It has long been known that plants emit and sense a wide variety of chemical signals; they also manufacture chemicals that deter predators and release others that have psychotropic effects for pollinators, encouraging them to revisit that particular plant again. As researchers continue to investigate the interplays between electrical and chemical signaling, gene structures, and plant behaviors, it becomes increasingly clear that, whatever one's position on the anthropocentrically laden word "intelligence," plants interpret a wide range of information about their environments and respond to challenges in remarkably nuanced and complex ways.

TECHNICAL COGNITION

Cognitive biology, along with related research in phytobiology discussed above, opens the concept of cognition to a broad compass, and to that extent, it is consistent with the path I want to pursue here. However, these research endeavors miss the opportunity to think beyond the biological to technical cognition, despite redefining terms in ways that partially enable that extension. To illustrate, I turn to the view of cognition proposed by Humberto R. Maturana and Francisco J. Varela in their seminal work *Autopoiesis and Cognition: the Realization of the Living* (1980). Maturana and Varela are distinct from the science of cognitive biology, associated instead with the Chilean School of Biology of Cognition; nevertheless, their views are close enough to cognitive biology to show the modifications necessary to extend cognition to technical systems.

Although they agreed about the cognitive capabilities of living organisms, they disagreed about whether these capabilities could be extended to technical systems—Maturana dissenting, Varela embracing. The disagreement is understandable, for their vision of what constituted cognition made the extension to technical systems far from obvious. In their view, cognition is intimately bound up with the recursive processes whereby an organism's organization determines its structures, and its structures determine its organization, in cycles of what Andy Clark (2008) subsequently called continuous reciprocal causality (note, however, that Maturana and Varela would not have used the term *causality* because an essential part of their vision was the closed or autopoietic nature of the living). Cognition, for them, is nothing other than this informational closure and the recursive dynamics it generates. Their postulated informational closure of organisms makes the extension to technical systems problematic, as technical systems are self-evidently *not* informationally closed but accept information inputs of various kinds and generate information outputs as well. Exploring more fully the cognitive capacities of technical systems, then, requires another definition of cognition than the one they adopted.

In *The Embodied Mind: Cognitive Science and Human Experience* (1991), Varela and coauthors Evan Thompson and Eleanor Rosch extend these ideas into comparisons between the cellular automata (a kind of computer simulation) and the emergence of cognition within biological cells (1991, 150–52). Their definition of enaction is consistent with the approach that I follow, insofar as it recognizes that cognition emerges from context-specific (i.e., embodied) interactions. "We propose as a name the term *enactive* to emphasize the growing conviction that cognition is not the representation of a pregiven world by a pregiven mind but is rather the enactment of a world and a mind on the basis of a history of the variety of actions that a being in the world performs. The enactive approach takes seriously, then, the philosophical critique of the idea that the mind is a mirror of nature but goes further by addressing this issue from within the heartland of science" (1991, 9).

In his later work, Varela was also interested not only in computer simulations but in creating autonomous agents within simulations, an approach known as Artificial Life (Varela and Bourgine 1992). Several years ago pioneers in this field argued that *life* is a theoretical program that can be instantiated in many different kinds of platforms, technological as well as biological (von Neumann 1966; Langton 1995;

Rosen 1991). For example, in an effort to show that technical systems could be designed to carry out biological functions, John von Neumann introduced the idea of "self-reproducing automata" (1966). More recently, John Conway's game of "life" (Gardner 1970) has often been interpreted as generating different kinds of species that can perpetuate themselves—as long as the computer does not malfunction or the electric current does not shut down. These caveats point to an insurmountable obstacle these researchers faced in arguing that life could exist in technical media, namely that such technical "life" can never be fully autonomous in its creation, maintenance, and reproduction. From the vantage of hindsight, I think this field of inquiry, although useful and productive in generating controversies and questions, was finally doomed to failure because technical systems can never be fully alive. But they *can* be fully cognitive. Their overlap with biological systems, in my view, should not be focused on "life itself" (as Rosen [1991] put it), but on cognition itself.

Following a path that has occupied me for several years, I offer a definition that will allow me to expand outward to include technical as well as biological cognition. *Cognition is a process that interprets information within contexts that connect it with meaning.* For me, the genesis of this formulation lay in Claude Shannon's theory of information (Shannon and Weaver 1948), in which he shifted the emphasis from a semantic basis for information to the selection of message elements from a set, for example, letters in an alphabet. This way of thinking about information has been enormously fruitful, as James Gleick has explained (2012), for it allowed the development of theorems and engineering practices that extended far beyond natural languages to information processes in general, including binary codes. From a humanities perspective, however, it had a major disadvantage. As Warren Weaver emphasized in his introduction to Shannon's classic work (Shannon and Weaver 1948), it appeared to sever information from meaning. Since the quest for meaning has always been central to the humanities, this meant that information theory would have limited usefulness for humanistic inquiries.

In retrospect, I think Weaver overstated the case in subtle but significant ways. As Shannon knew quite well, the process of selection, which he expressed as a function of probabilities, is not entirely divorced from a message's content and consequently from its meaning. In fact, the conditional probabilities of what message elements will follow their predecessors are already partially determined by the distribution of letters and their relative frequencies within a given language. In English and Romance languages, for example, there is a nearly 100 percent chance that a "q" will be followed by a "u," a higher than random chance that an "e" will be followed by a "d," and so forth. Shannon (1993) linked this idea to the redundancy of English (and other languages), and the theorems that followed were crucial for information compression techniques still in use for telephonic and other kinds of communication transmissions.

Nevertheless, to arrive at meaning, the constraints operating through selection processes are not enough. Something else is needed: context. Obviously, the same sentence, uttered in different circumstances, can change its meaning completely. The missing link between Shannon's view of information and context was supplied for me in a seminar given by the theoretical physicist Edward Fredkin, when he casually observed, "The meaning of information is given by the processes that interpret it" (Hayles 2012, 150). Although Fredkin gave no indication he thought this idea was particularly powerful, it hit me like a bolt of lightning. It blows the problem of meaning wide open, for processes occur within contexts, and *context* can be understood in radically diverse ways for different situations. It applies to utterances of natural language between humans, but it equally well describes the informational processes by which plants respond to information embedded in the chemicals they absorb, the behavior of octopi when they sense potential mates in their vicinity, and the communications between layers of code in computational media. In another context, the insight can also be related to how the brain processes sensory information, in which action potentials and patterns of neural activity may be experienced in different ways depending on which part of the brain engages them (see for example chapter 21, "Sensory Coding," in Kandel and Schwartz 2012, 449–74).[5]

Consistent with Fredkin's explosive insight is the processual and qualitative view of information (as distinct from the quantitative theory developed by Shannon) proposed by the French "mechanologist" Gilbert Simondon in the 1960s as part of his overarching philosophy focusing on processes rather than hylomorphic concepts (form and matter). For Simondon, reality itself is the tendency to engage in processes. A central metaphor for him is the concept of potential energy always tending to flow from a higher state to a lower one, but never coming to a stable equilibrium, only transitional metastable states. He called this flow "information" and thought it is inherently connected with meaning (Simondon 1989; see also Scott 2014; Iliadis 2013; and Terranova 2006). Similar to Fredkin's insight, information in this view is not a statistical distribution of message elements but the result of embodied processes emerging from an organism's embeddedness within an environment. In this sense, the processes that nonconscious cognition uses to discern patterns are constantly in motion, reaching metastable states as patterns are discerned and further reinforced when temporal matching with the reverberations between neural circuits cause them to be fed forward to consciousness. These processes of discerning patterns are always subject to new inputs and continuing transformations as the nonconscious and conscious contexts in which they are interpreted shift from moment to moment. In Simondon's terms, the transfer from one neural mode of organization to another can be conceived as a transfer from one kind of potential energy to another. The information coming to consciousness has already been laden with meaning (that is, interpreted in the relevant contexts) by the cognitive nonconscious; it achieves further meaning when it is re-represented within consciousness.

As we will see in chapter 5, interpretation within contexts also applies to the nonconscious cognitive processes of technical devices. Medical diagnostic systems, automated satellite imagery identification, ship navigation systems, weather prediction programs, and a host of other nonconscious cognitive devices interpret ambiguous or conflicting information to arrive at conclusions that rarely if ever are completely certain. Something of this kind also happens with the cognitive nonconscious in humans. Integrating multiple somatic markers, it too must synthesize conflicting and/or ambiguous information to arrive at interpretations that may feed forward into consciousness, emerging as emotions, feelings, and other kinds of awareness upon which further interpretive activities take place.

In automated technical systems, nonconscious cognitions are increasingly embedded in complex systems in which low-level interpretative processes are connected to a wide variety of sensors, and these processes in turn are integrated with higher-level systems that use recursive loops to perform more sophisticated cognitive activities such as drawing inferences, developing proclivities, and making decisions that feed forward into actuators, which perform actions in the world.

In an important sense, *these multi-level systems represent externaliza- tions of human cognitive processes.* Although the material bases for their operations differ significantly from the analogue chemical/electrical signaling in biological bodies, the *kinds* of processes have similar in- formational architectures. In addition, technical systems have the ad- vantage of working nonstop 24/7, something no biological body can do, and of processing vast amounts of information much faster than humans can. It should not be surprising that human and technical nonconscious cognitions share attributes in common, because brains (deploying nonconscious cognition in their own operations) designed them.

PARSING COGNITION

With this background, let us return to parse my definition more fully, since it is foundational for the arguments to follow. *Cognition is a pro- cess:* this implies that cognition is not an attribute, such as intelli- gence is sometimes considered to be, but rather a dynamic unfolding within an environment in which its activity makes a difference. For example, a computer algorithm, written as instructions on paper, is not itself cognitive, for it becomes a process only when instantiated in a platform capable of understanding the instruction set and carry- ing it out. *That interprets information:* interpretation implies a choice. There must be more than one option for interpretation to operate. In computational media, the choice may be as simple as the answer to a binary question: one or zero, yes or no. Other examples include, in the C++ programming language, commands such as "if" and "else" state- ments ("if" indicates that a procedure should be implemented only if certain conditions are true; "else" indicates that if these conditions are not met, other procedures should be followed). Moreover, these commands may be nested inside each other to create quite complex decision trees. Choice here, of course, does not imply "free will" but rather programmatic decisions among alternative courses of action, much as a tree moving its leaves to maximize sunlight does not imply free will but rather the implementation of behaviors programmed into the genetic code.

In *Cognitive Biology*, Gennaro Auletta (2011) writes that "biological systems represent the integration of the three basic systems that are involved in *any* physical process of information-acquiring: The processor, the regulator, and the decider" (200). In unicellular organisms, the "decider" may be as simple as the lipid membrane that "decides" which chemicals to admit and which to resist. In more complex multicellular organisms such as mammals and in networked and programmable media, the interpretive possibilities grow progressively more multileveled and open-ended. *In contexts that connect it with meaning:* the implication is that meaning is not an absolute but evolves in relation to specific contexts in which interpretations performed by the cognitive processes lead to outcomes relevant to the situation at that moment. Note that context *includes* embodiment. Lest I be misunderstood, let me emphasize that technical systems have completely different instantiations than biological life-forms, which are not only embodied but also embedded within milieus quite different from those of technical systems.[6] These differences notwithstanding, both technical and biological systems engage in meaning-making within their relevant instantiated/embodied/embedded contexts. For high-level cognitive processes such as human thought, the relevant contexts may be very broad and highly abstract, from deciding whether a mathematical proof is valid to questioning if life is worth living. For lower-level cognitive processes, the information may be the sun's angle for trees and plants, the location of a predator as a school of minnows darts to evade it, or the modulation of a radio beam by a radio-frequency identification (RFID) chip that encodes it with information and bounces it back. In this framework, all these activities, and millions more, count as cognitive.

A meta-implication is that humans do not have a lock on which contexts and levels are able to generate meanings. Many technical systems, for example, operate through communication signals such as radio waves, microwaves, and other portions of the electromagnetic spectrum inaccessible to direct human perception. To unaided human senses, the signals bouncing around the atmosphere are both imperceptible and meaningless, but to technical devices that operate in contexts relevant to them, they are filled with meaning. Traditionally, the humanities have been concerned with meanings relevant to humans in human-dominated contexts. The framework developed here challenges that orientation, insisting cognitive processes happen within a broad spectrum of possibilities that include nonhuman animals and plants as well as technical systems. Moreover, the meanings generated within these contexts, deeply worthy of consideration in their own right, are also consequential for human outcomes as well, from the flourishing of trees in rain forests to the communication signals emanating from a control tower to aircraft within its purview. This framework emphasizes that these different kinds of meanings are entangled together in ways that transcend any single human viewpoint and that cannot be bounded by human interests alone. As our view of what counts as cognition expands, so too do the realms in which interpretations and meanings emerge and evolve. All of these, this framework implies, count as meaning making and consequently should be of potential interest to the humanities, as well as to the social and natural sciences.

THE TRIPARTITE FRAMEWORK OF (HUMAN) COGNITION

Turning now specifically to human cognition, I develop this view with a tripartite framework that may be envisioned as a pyramid with three distinct layers (fig. 1, p. 40). At the top are consciousness and unconsciousness, grouped together as modes of awareness. As noted earlier, research on the "new" unconscious sees it as a kind of broad environmental scanning in which events are heeded and, when appropriate, fed forward to consciousness (Hassin, Uleman, and Bargh 2005). The new unconscious differs from the psychoanalytic unconscious of Freud and Lacan in that it is in continuous and easy communication with consciousness. In this view the psychoanalytic unconscious may be considered as a subset of the new unconscious, formed when some kind of trauma intervenes to disrupt communication and wall off that portion of the psyche from direct conscious access. Nevertheless, the psychoanalytic unconscious still expresses itself to consciousness through symptoms and dreams susceptible to psychoanalytic interpretation. The modes of awareness, designating the neurological functions of consciousness and the communicating unconscious, form the top layer of the pyramid.

The second part of the tripartite framework is nonconscious cognition, described in detail elsewhere (Hayles 2012). Unlike the unconscious, it is inherently inaccessible to consciousness, although its outputs may be forwarded to consciousness through reverberating circuits (Kouider and Dehaene 2007). Nonconscious cognition integrates somatic markers such as chemical and electrical signals into coherent body representations (Damasio 2000; Edelman 1987). It also integrates sensory inputs so that they are consistent with a coherent view of space and time (Eagleman 2012). In addition, it comes online much faster than consciousness and processes information too dense, subtle, and noisy for consciousness to comprehend. It discerns patterns that consciousness is unable to detect and draws inferences from them; it anticipates future events based on these inferences; and it influences behavior in ways consistent with its inferences (Lewicki, Hill, and Czyzewska 1992). No doubt nonconscious cognition in humans evolved first, and consciousness and the unconscious were subsequently built on top. Removed from the confabulations of conscious narration, nonconscious cognition is closer to what is actually happening in the body and the outside world; in this sense, it is more in touch with reality than is consciousness. It comprises the broad middle layer of the tripartite framework.

The even broader bottom layer comprises material processes. Although these processes are not in themselves cognitive, they are the dynamic actions through which all cognitive activities emerge. The crucial distinguishing characteristics of cognition that separate it from these underlying processes are choice and decision, and thus possibilities for interpretation and meaning. A glacier, for example, cannot choose whether to slide into a shady valley as opposed to a sunny plain. In contrast, as Auletta explains, "any biological system . . . produces variability as a response to environmental challenges and tries to integrate [these] aspects inside itself" (2011, 200). In general, material processes may be understood through the sum total of forces acting upon them. A special case is formed by criticality phenomena, structured so that even minute changes in initial conditions may change how the system evolves. Even here, the systems remain deterministic, although they are no longer predictable. There are many examples of material processes that can self-organize, such as the Belousov-Zhabotinsky (BZ) inorganic reaction. However, there remain crucial distinctions between such far-from-equilibrium systems and living organisms, for whom choices, decisions, and interpretations are possible. As Auletta points out, "biological systems are more than simply dissipative self-organizing systems, for the reason that they can negotiate a changing or nonstationary environment in a way that allows them to endure (to change in an adaptive sense) over substantial periods of time" (2011, 200). Material processes may however be harnessed to perform cognitive functions when natural or artificial constraints are applied in such a way as to introduce choice and agency into the system (Lem 2014), for example, through the interactions of multiple independent agents in complex environments.

Although the pyramidal shape of the tripartite framework may seem to privilege the modes of awareness over nonconscious cognitions and material processes, inasmuch as they occupy the top strata, a countervailing force is expressed through the pyramid volumes. The modes of awareness, precisely because they come at the top, reign over the smallest volume, a representation consistent with the roles they play in human psychic life. Nonconscious cognition occupies a much greater volume, consistent with the processes it performs as the neurological function mediating between the frontal cortex and the rest of the body. Material processes occupy a vast volume, consistent with their foundational role from which all cognition emerges.

Although the tripartite framework divides human processes into three distinct layers for analytical clarity, in reality complex recursive loops operate throughout the system to connect the layers to each other and connect different parts of each layer within itself. Each layer operates dynamically to influence the others all the time, so the system is perhaps better described as a dynamic heterarchy rather than a linear hierarchy, a view that animates and interconnects the system as it evolves in real time. Consequently, the structure sketched above is a first approximation. It is not so much meant to settle questions as to catalyze boundary issues and stimulate debates about how the layers interact with each other. That said, it nevertheless serves as a starting point to discuss issues of agency and to distinguish between actors and agents.

Because cognition in this framework is understood as inseparable from choice, meaning, and interpretation, it bestows special functionalities not present in material processes as such. These include flexibility, adaptability, and evolvability. Flexibility implies the ability of an organism or technical system to act in ways responsive to changing conditions in its environment. Whereas a ball thrown toward a window has no choice to alter its trajectory, a self-driving car can respond with a large repertoire of possibilities to avoid damage. As indicated above, flexibility is present in all living organisms to some extent, even those lacking central nervous systems.[7] Adaptability denotes developing capacities in response to environmental conditions. Examples include changed neurological functioning in plants, animals, and humans in response to environmental stresses or opportunities, such as the neurological changes human brains undergo through extensive interactions with digital media (Hayles 2012). Evolvability is the possibility to change the programming, genetic or technical, that determines the repertoire of responses. Genetic and evolutionary algorithms are examples of technical systems with these capabilities (Koza 1992), as are computers that can reconfigure their own firmware, rearranging logic gates to solve problems with maximum efficiency (Ling 2010). Biological examples are of course everywhere, as biologists from Darwin and Wallace on have confirmed. The important point is that material processes do not possess these capabilities in themselves, although they may serve to enhance and enlarge cognitive capabilities when enrolled as supports in an extended cognitive system.

ACTORS AND AGENTS

It is fashionable nowadays to talk about a human/nonhuman binary, often in discourses that want to emphasize the agency and importance of nonhuman species and material forces (Bennett 2010; Grosz 2011; Braidotti 2013). To my mind, there is something weird about this binary. On one side are some seven billion individuals, members of the Homo sapiens species; on the other side sits everything else on the planet, including all the other species in the world, and all the objects ranging from rocks to clouds. This binary, despite the intentions of those who use it, inadvertently reinstalls human privilege in the vastly disproportionate weight it gives to humans. Some theorists in the ecological movement are developing a vocabulary that partially corrects this distortion by referring to the "more-than-human" (Smith 2011), but the implicit equivalence of the human world to everything else still lingers.[8]

Recognizing that binaries can facilitate analysis (their limitations notwithstanding), I propose another distinction to replace human/nonhuman: *cognizers versus noncognizers.* On one side are humans and all other biological life forms, as well as many technical systems; on the other, material processes and inanimate objects. At the very least, this distinction is more balanced in the relative weights it gives to the two sides than the very unbalanced human/nonhuman formulation. This binary (like all binaries) is not innocent of embedded implications. In particular, it foregrounds cognition as a primary analytical category. Skeptics may object that it too reinstalls human privilege, since humans have higher and more extensive cognitions than other species. However, this binary is part of a larger cognitive ecology emphasizing that *all* life forms have cognitive capabilities, including some that exceed human cognitions (smell in dogs, for example).

Moreover, because only cognizers can exercise choice and make decisions, they have special roles to play in our current environmental crises and the sixth mass extinction already underway. The one motivation that all life-forms share is the struggle to survive. As environmental stresses increase differentially, cognizers at all levels, from worms to humans, will make choices that tend to maximize their chances for survival. Admittedly, species with higher cognitive capabilities can supervene this motivation as it interacts with other priorities—as many humans are doing at present. Having an analytical category that emphasizes choice may help to foreground our common causes with other cognizers and draw our attention more vividly to the fact that we all make choices, and that these choices matter, individually and collectively. Moreover, the capabilities that cognition bestows—flexibility, adaptability, evolvability—imply that cognizers have special roles to play in our evolving planetary ecologies. Finally, this framework sets up the possibility that cognitive technologies may perform as ethical actors in the assemblages they form with biological life-forms, including humans.

For their part, noncognizers may possess agential powers that dwarf anything humans can produce: think of the awesome powers of an avalanche, tsunami, tornado, blizzard, sandstorm, hurricane. Faced with these events, humans utterly lack the ability to control them; the best they can do is get out of the way. Moreover, since material processes are the underlying forces that nourish and give rise to life, they deserve recognition and respect in their own right, as foundational to everything else that exists (Strang 2014). What they cannot do, acting by themselves, is make choices and perform interpretations. A tornado cannot choose to plow through a field rather than devastate a town. Material processes, of course, respond to contexts and, in responding, change them. But because they lack the capacity for choice, they perform as agents, not as actors embedded in cognitive assemblages with moral and ethical implications.

I propose a further shift in terminology that clarifies the different roles performed by material processes and nonconscious cognizers. I suggest reserving the term *actors* for cognizers, and *agents* for material forces and objects. This latter category includes objects that may act as cognitive supports; it also includes material forces that may be harnessed to perform cognitive tasks when suitable constraints are introduced, for example, when electrical voltages are transformed into a bit stream within a computational medium.

Fueled by global capitalism, technical cognitive systems are being created with ever more autonomy, even as they become increasingly pervasive within developed societies. As David Berry (2015) among others points out, there is no technical agency without humans, who design and build the systems, supply them with power and maintain them, and dispose of them when they become obsolete. Nevertheless, the pockets within which technical systems operate autonomously are growing larger and more numerous. Examples include environmental monitoring systems, surveillance and communication satellites, digital search engines, and language learning systems, among many others. Perhaps an appropriate way to think about the growing autonomy of these systems is as punctuated agency, analogous to "punctuated equilibrium" (Gould 2007). Like punctuated equilibrium, punctuated agency operates within regimes of uneven activity, longer periods when human agency is crucial, and shorter intervals when the systems are set in motion and proceed on their own without direct human intervention.

Even within the autonomous regions, the effects of technical cognitions are not contained wholly within the technical systems. They interact with human complex systems to affect myriad aspects of human and biological life. In this respect, even the cognizer/noncognizer binary falls short because it fails to capture the powerful and subtle ways in which human and technical cognizers interact with each other as well as with noncognizing objects and material forces. Water is a good example (Strang 2014): on its own it exercises agency through such phenomena as waterfalls, rain, snow, and ice; incorporated into biological bodies, it provides fluids essential for life; run through a turbine, it contributes to the cognitions and effectiveness of a computerized hydroelectric power system. To express more adequately the complexities and pervasiveness of these interactions, we should resist formulations that reify borders and create airtight categories. The better formulation, in my view, is not a binary at all but interpenetration, continual and pervasive interactions that flow through, within, and beyond the humans, nonhumans, cognizers, noncognizers, and material processes that make up our world.

WHY COMPUTATIONAL MEDIA ARE
NOT JUST ANOTHER TECHNOLOGY

In *What Technology Wants,* Kevin Kelly (2010) argues that technologies develop along trajectories that he anthropocentrically identifies with "desire," including ubiquity, diversity, and intensity. As the provocation of his title indicates, his discussion fails to give a robust account of how human agency enters this picture. Nevertheless, there is a kernel of insight here, which I rephrase as this: technologies develop within complex ecologies, and their trajectories follow paths that optimize their advantages within their ecological niches. The advent of photography in the mid-to-late nineteenth century, for example, preempted the category of landscape description, and consequently literary novels readjusted their techniques, moving away from the pages of landscape description notable in late-eighteenth- and early-nineteenth-century novels and into stream of consciousness strategies, an area that photography could not exploit as effectively. As Cynthia Sundberg Wall has shown (2014, esp. chapters 1–3, 2–95), literary descriptive techniques are enmeshed within a cultural matrix of techniques of vision, including microscopes, telescopes, maps, and architectural diagrams. The dynamics of competition, cooperation, and simulation between media forms are powerful analytics for understanding technological change (Fuller 2007; Hansen 2015; Gitelman 2014).

In these terms, computational media have a distinct advantage over every other technology ever invented. They are not necessarily the most important for human life; one could argue that water treatment plants and sanitation facilities are more important. They are not necessarily the most transformative; that honor might go instead to transportation technologies, from dirt roads to jet aircraft. Computational media are distinct, however, because they have a *stronger evolutionary potential* than any other technology, and they have this potential because of their cognitive capabilities, which among other functionalities, enable them to simulate any other system.

We may draw an analogy with the human species. Humans are not the largest life-form; they are not the strongest or the fastest. The advantages that have enabled them to achieve planetary dominance within their ecological niche are their superior cognitive capabilities. Of course, we are long past the era when the Baconian imperative for humans to dominate the earth can be embraced as an unambiguous good. In an era of ecological crises, global warming, species extinction, and similar phenomena, the advent of the Anthropocene, in which human influences are changing geological and planetary records, is properly cause for deep concern and concerted political activism around climate change, preservation of habitats, and related issues.

The analogy with the cognitive capacities of computational media suggests that a similar trajectory of worldwide influence is now taking place within technical milieus. Fueled by the relentless innovations of global capital, computational media are spreading into every other technology because of the strong evolutionary advantages bestowed by their cognitive capabilities, including water treatment plants and transportation technologies but also home appliances, watches, eyeglasses, and everything else, investing them with "smart" capabilities that are rapidly transforming technological infrastructures throughout the world. Consequently, technologies that do not include computational components are becoming increasingly rare. Computational media, then, are not just another technology. They are the quintessentially *cognitive* technology, and for this reason have special relationships with the quintessentially cognitive species, *Homo sapiens.*

Note that this position should not be conflated with technological determinism. As Raymond Williams has astutely observed, such evolutionary potentials operate within complex social milieus in which many factors operate and many outcomes are possible: "We have to think of determination not as a single force, or a single abstraction of forces, but as a process in which real determining factors—the distribution of power or of capital, social and physical inheritance, relations of scale and size between groups—set limits and exert pressures, but neither wholly control nor wholly predict the outcome of complex activity within or at these limits, and under or against these pressures" (Williams 2003, 13). In fact, one can argue that the larger the cognitive components of technological systems, the more unpredictable are their specific developments, precisely because of the qualities conferred by cognition, namely flexibility, adaptability, and evolvability. As global capital continues to innovate ways in which computational media may be infused into other technologies, the e-waste created by their exponential growth increasingly poisons environments where they end up, disproportionately, in poor, underprivileged, and underfunded countries. Given that the cognitive capabilities of technical media are achieved at considerable cultural, social, political, and environmental costs, we can no longer avoid the ethical and moral implications involved in their production and use.

TECHNOLOGICAL COGNITION AND ETHICS

As we have seen, *choice* in my framework has a very different meaning than in ethical theories, where it is associated with free will. What ethical approaches are appropriate to the former, which I will call CHOICEII (interpretation of information), as distinct from CHOICEFW (free will)? Bruno Latour (1992) touches on this question when he suggests that the "missing masses" of ethical actors (by analogy with the missing mass/energy that physicists need to explain the universe's inflation) are technical artifacts: "here they are, the hidden and despised social masses who make up our morality" (1992, 227). Using simple examples of seat belts and hydraulic door closers, Latour shows that technical artifacts encourage moral behavior (annoying buzzers that remind drivers to fasten seat belts) and influence human habits (speed bumps influencing drivers not to speed in school zones) (2002). In these examples, the technical objects are either passive or minimally cognitive. Even at this modest level, however, artifacts act as "mediators" influencing human behaviors, notwithstanding that they often sink into the background and are perceived unconsciously (Latour 1999, 2002; Verbeek 2011).

When artifacts embody higher levels of cognition, they can intervene in more significant and visible ways. Peter-Paul Verbeek develops a philosophical basis for thinking about technical systems as moral actors and suggests how to design technologies for moral purposes (2011, 135). The Fitbit bracelet (my example, not his) encourages fitness by monitoring heart rate, keeping track of workouts, noting calories burned, and measuring distances covered and stairs climbed. None of these devices absolutely compel obedience, as Latour acknowledges, because there are always ways to defeat their behavioral intent. Nevertheless, they have cumulative (and expanding) effects that significantly affect human social behaviors and unconscious actions.

Following Latour's lead in thinking about technical systems as "mediators," Verbeek develops the argument further by showing how technologies such as obstetric ultrasound not only open new areas for ethical consideration (for example, whether to abort a malformed or, even more distressing, a female fetus) but also reconfigure human entities in new ways (the fetus becoming a medical patient viewable by the physician). In the entangled web of human and technical actors, Verbeek argues, both humans and technics share moral agency and, implicitly, moral responsibility: "moral agency is distributed among humans and nonhumans; moral actions and decisions are the products of human-technology associations" (Verbeek 2011, 53).[9]

Like Verbeek, Latour emphasizes the unexpected effects of technological innovations, arguing that technological systems almost always modify and transform the ends envisioned in their original designs, opening up new possibilities and, in the process, entangling means and ends together so that they can no longer reasonably be regarded as separate categories.[10] The thrust of this argument, of course, is to defuse the objection that technological artifacts are merely the means for ends established by humans. Examples of technologies invented for one purpose and reappropriated for another are legion, from the typewriter, initially invented for blind people, to the Internet, originally intended as a place where scientific researchers could exchange results.

While Latour and Verbeek offer valuable guidance, to my mind their arguments do not go far enough. With technologies capable of significant decision making—for example, autonomous drones—it does not seem sufficient to call them "mediators," for they perform as actors in situations with ethical and moral consequences. One might argue, as Verbeek does, that distributed agency implies distributed responsibility, but this raises the prospect of a technological artifact being called to account for performing the actions programmed into it, a misplaced ethical judgment reminiscent of medieval animal trials in which starlings were executed for chattering in church and a pig was hanged for eating a communion wafer.

Ethical theories, for their part, are often intensely anthropocentric, focusing on individual humans as the responsible agents to whom ethical standards should apply, as in Emmanuel Levinas's complex notion of the Other's face (1998). Although some theories extend this to animals (for example, Tom Regan's suggestion [2004] that mammals over a certain age should be considered subjects of a life and therefore have ethical rights), few discuss the role of technical cognizers as responsible technical actors. Latour is certainly right to point to human-technical assemblages as transformative entities that affect ends as well as means, but he offers little guidance on how to assess the ethical implications of such assemblages. If, to use Latour's example, neither guns nor people are the agents responsible for gun violence but rather the gun-person collective they form (Latour 1999, 193), surely drone-with-pilot is a much more potent assemblage than either by itself/himself.

To assess such assemblages, we should move from thinking about the individual and CHOICEFW as the focus for ethical or moral judgment, and shift instead to thinking about CHOICEII and the consequences of the actions the assemblage as a whole performs. Jeremy Bentham suggested a similar move when he wrote, "The general tendency of an act is more or less pernicious according to the sum total of its consequences, i. e., according to the difference between the sum of its good consequences and the sum of its bad ones" ([1780] PDF, 43). We need not subscribe to all the tenets of utilitarianism to accept this as an adequate framework in which to evaluate the effects of cognitive assemblages that include technical actors. Drone pilots cannot be considered simply as evil for killing other humans; even less so can the drone itself. Rather, they act within structured situations that include tactical commanders, lawyers, and presidential staff, forming assemblages in which technological actors perform constitutive and transformative roles along with humans. The results should therefore be evaluated *systemically* in ways that recognize not all of the important actors are human, an argument developed further in part 2. Moreover, drone assemblages are part of larger conflicts that includes suicide bombers, IEDs, military incursions, insurgent resistance, and other factors. The cognitive assemblages in such conflicts are differentially empowered by the kinds of technologies they employ as well as by how the humans enmeshed within them act. The consequences of the assemblages further interact with existing discourses and ethical theories in dynamic, constantly shifting constellations of opposing interests, sovereign investments, personal decisions, and technological affordances. Attempting to evaluate moral and ethical effects from the actions of individual people alone by focusing on CHOICEFW is simply not adequate to assess the complexities involved. As part 2 argues more fully, we need frameworks that explore the ways in which the technologies interact with and transform the very terms in which ethical and moral decisions are formulated.

We can see the inadequacy of remaining within individual-focused frameworks by considering the justification for designing robot weapons offered by Ronald C. Arkin, Regents Professor of computer science at Georgia Tech, compared with the drone theory of Grégoire Chamayou. Arkin, who has Defense Advanced Research Projects Agency (DARPA) grants to develop autonomous robot warriors for the battlefield, argues that robots may be morally superior to human warriors because they would be forbidden by their programming to commit atrocities, immune to emotional stress and the bad decisions that can accompany it, and able to direct their lethal encounters more precisely, minimizing collateral damage (Arkin 2010, 332–41). His critics attack these claims on a number of fronts; perhaps the most compelling is the objection that once robot warriors are available, they would likely be used more widely and indiscriminately than human warriors, where the prospect of putting one's troops "in harm's way" acts as a significant restraint on military and political leaders.

Evaluating the claims for robot morality requires a larger interpretive frame than the one Arkin uses. Leaving aside the question of whether robots would in fact be programmed to follow the rules of war established by international treaties (and whether these rules could ever make war "moral," an issue explored in part 2), I note that he treats the robots in the same terms as human individuals (but equipped with better sensors and decision-making capabilities) rather than as technical systems embedded in complex human-technical assemblages.

Grégoire Chamayou (2015) is subtler in interrogating how the specific rules of engagement for drone pilots cause conventional standards of appropriate behavior in warfare to be transformed and reinterpreted to accommodate the pilots' actions. For example, he points out that traditional accounts of war distinguish sharply between soldiers and assassins. Whereas the former are considered honorable because, by entering a field of combat, they establish who is an enemy combatant and also put their own bodies at risk, assassins are cowardly because they may strike targets who are not combatants and do not necessarily put themselves at risk in doing so. Applied to drone pilots, these views could force them to be counted as assassins rather than soldiers. To mitigate the situation, the US military has emphasized that drone pilots may be suffering from post-traumatic stress disorder and in this sense are putting themselves at risk as well. Although Chamayou has his own agenda and often is one-sided in his appraisals (as argued in part 2), his analyses nevertheless show that the consequences of human-technical assemblages include not only the immediate results of actions but also far-reaching transformations in discourses, justifications, and ethical standards that attempt to integrate those actions into existing evaluative frameworks.

The more powerful the cognitive capabilities of technical systems, the more far-reaching are the results and transformations associated with them. Drones are especially controversial examples, but technical cognitive systems employing CHOICEII are all around us and operating largely under the radar of the general public, including expert medical systems, automated trading algorithms, sensing and actuating traffic networks, and surveillance technologies of all kinds, to mention only a few. To analyze and evaluate their effects, we need robust frameworks that recognize technical cognition as a fact, allowing us to break out of the centuries-old traditions that identify cognition solely with (human) consciousness. We also need a more accurate picture of how human cognitive ecology works, including its differences from and similarities to technical cognition. Finally, we need a clear understanding of how cognizers differ from material processes, which includes a definition of cognition that sets a low threshold for participation but includes ways to scale up to much more sophisticated cognitions in humans, nonhuman life forms, and technical systems. Added together, these innovations amount to nothing less than a paradigm shift in how we think about human cognition in relation to planetary cognitive ecologies, how we analyze the operations and ethical implications of human-technical assemblages, and how we imagine the role that the humanities can and should play in assessing these effects.

In conclusion, let me address the role of humanistic critique. If thought in general is associated with consciousness, critique is even more so. Some may object that challenging the centrality of reason in cognitive processes undermines the nature of critique itself. Yet consciousness alone cannot explain why scholars choose certain objects for their critique and not others, nor can it fully address the embodied and embedded resources that humanities scholars bring to bear in their rhetorical, analytical, political, and cultural analyses of contemporary issues. Without necessarily realizing it, humanities scholars

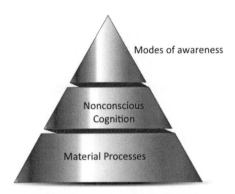

Figure 1. The tripartite framework of (human) cognition as a pyramid

have always drawn upon the full resources of human cognitive ecologies (fig. 1), both within themselves and within their interlocutors. Recognizing the complexities of these interactions does not disable critique; on the contrary, it opens critique to a more inclusive and powerful set of resources with which to analyze the contemporary situations that confront us, including but not limited to the entanglements and interpenetrations of human and technical cognitive systems. That is the importance, and the challenge, of the cognitive nonconscious to the humanities today.

CHAPTER 2

Interplays between Nonconscious Cognition and Consciousness

Whereas chapter 1 explored cognition across the full spectrum of biological life-forms, this chapter focuses on the relation of nonconscious cognition to consciousness, specifically in humans. It explores the ongoing reassessments of consciousness in cognitive science, human neurology, and related fields in view of the important functions that nonconscious cognitions perform. It discusses theories about how nonconscious cognitive processes relate to consciousness, the mechanisms that enable nonconscious processes to feed forward intuitions to conscious awareness, and the important role that temporality plays in these mechanisms. The chapter concludes with a consideration of how recognizing the importance of nonconscious cognition can inform contemporary debates about consciousness. It also discusses how nonconscious processes have been represented in non-Western traditions, specifically meditative techniques and mindfulness, and how philosophical movements such as speculative realism overlap with, and diverge from, the framework presented here.

COSTS OF CONSCIOUSNESS

As noted in chapter 1, we can distinguish between core or primary consciousness, widely shared among mammals and other life-forms, and higher or secondary consciousness, unique to humans and some primates. According to Thomas Metzinger (2004, 107–305), a contemporary German philosopher, core consciousness creates a mental model of itself that he calls a "Phenomenal Self-Model" (PSM) (107); it also creates a model of its relations to others, the "Phenomenal Model of the Intentionality Relation" (PMIR) (301–05). Neither of these models could exist without consciousness, since they require the memory of past events and the anticipation of future ones. From these models, the experience of a self arises, the feeling of an "I" that persists through time and has a more or less continuous identity. The PMIR allows the self to operate contextually with others with whom it constructs an intentionality relation.

The sense of self, Metzinger argues, is an illusion, facilitated by the fact that the construction of the PMS and the PMIR models are transparent to the self (that is, the self does not recognize them as models but takes them as actually existing entities). This leads Metzinger to conclude, "nobody ever was or had a self" (1). In effect, by positioning the self as epiphenomenal, he reduces the phenomenal experience of self back to the underlying material processes. In my view, we need not accept his claim that the self is an illusion to find the PMS and PMIR useful as explanations for how a sense of self evolves and operates. Philosopher of consciousness Owen Flanagan, following William James, tracks a similar line of reasoning: "the self is a construct, a model, a product of conceiving of our organically connected mental life in a certain way" (Flanagan 1993, 177). Who thinks the thoughts that we associate with the self? According to Flanagan (and James), the thoughts think themselves, each carrying along with it the memories, feelings, and conclusions of its predecessor while bearing them toward its successor.

Antonio Damasio (2000) holds a somewhat similar view, in the sense that he considers the self to be a construct created through experiences, emotions, and feelings a child has as she grows, rather than an essential attribute or possession. Damasio, however, also thinks that the self evolved because it has a functional purpose, namely to create a concern for preservation and well-being that propels the organism into action and thus guarantees "that proper attention is paid to the matters of individual life" (303). Owen Flanagan agrees: consciousness and the sense of self have functions, including serving as a clearing house of sorts where past experiences are recalled as memories and future anticipations are generated and compared with memories to arrive at projections and outcomes. In Daniel Dennett's metaphor (1992, 139–70), consciousness and the working memory it enables constitute the "workspace" (about which we will hear more later) where past, present, and future are put together to form meaningful sequences (256–66).

Meaning, then, can be understood at the level of core consciousness as an emergent result of the relation between the PMS and the PMIR—that is, between the self-model and models the self constructs of objects that it has an "intention toward." Damasio puts it more strongly; *there is no self without awareness of and engagement with others* (2000, 194, emphasis added). The self thus requires core consciousness, which constructs the PMS and the PMIR; without consciousness, a self could not exist. In humans (and some primates), the core self is overlaid with a higher-level consciousness capable of meta-level reasoning, including interrogations of meanings that call for interpretations. Through becoming aware of and reflecting on the self as a self, higher consciousness has the effect of reifying the sense of self even more. Higher consciousness, because it generates the verbal monologues that interpret the actions of the self, has a tendency also to become imperialistic, to appropriate to itself the entirety of consciousness and even of cognition, about which we will hear more in chapter 4. John Bickle neatly summarizes the situation when he discusses what he calls the Elaborate Practical Reasoning, author of the verbal narratives mentioned above. "What kind of self do [these narratives] create and express? Clearly, they create and express a causally efficacious self-image: one of a self not only in causal control of important cognitive, conscious, and behavior events but also aware of exerting this control" (Bickle 2003, 199). As he points out, however, this sense of control is largely illusory. "The inner narratives that create and sustain our selves are relatively impotent over information processes in other neural regions . . . [the] language systems summarize and then broadcast a highly edited snapshot of the outputs of the brain's cognitive processing networks" (201).

Nevertheless, this editing does have a strong adaptive purpose, for it creates and maintains a coherent picture of the world. As Gerald Edelman and Giulio Tononi put it, "Many neuropsychological disorders demonstrate that consciousness can bend or shrink, and at times even split but it does not tolerate breaks of coherence" (Edelman and Tononi 2000, 27). We can easily see how this quality would have advantages. Creating coherence enables the self to model causal interactions reliably, make reasonable anticipations, and smooth over the gaps and breaks that phenomenal experiences present. If a car is momentarily hidden by a truck and then reappears, consciousness recognizes this as the same car, often at a level below focused attention.

Inextricably woven with these advantages are costs, for safeguarding coherence above all frequently causes consciousness to misrepresent anomalous or strange situations. A number of experiments in cognitive psychology confirm this fact (not to mention the entire history of stage magic). In one now-famous situation (see video at http://www.youtube.com/watch?v=vJG698U2Mvo), subjects are shown a video of players passing a basketball and are asked to keep track of the passes. In the middle of the scene, someone dressed in a gorilla suit walks across the playing area, but a majority of subjects report that they saw nothing unusual (Simons and Chabis 2011, 8; Simons and Chabis 1999). In another staged situation, a man stops a passerby and asks for directions (Simons and Chabis 2011, 59). While the subject is speaking, two workmen carrying a vertical sheet of wood pass between them, momentarily blocking the view. After they have passed, the interlocutor has been replaced by another person (who has been walking unseen behind the wood panel), but the majority of subjects do not notice the discrepancy. Useful as is the tendency of consciousness to insist on coherence, these experiments show that one cost is the screening out of highly unusual events. Without our being aware of it, consciousness edits events to make them conform to customary expectations, a function that makes eyewitness testimony notoriously unreliable. Even in the most ordinary circumstances, consciousness confabulates more or less continuously, smoothing out the world to fit our expectations and screening from us the world's capacity for infinite surprise.

A second cost is the fact that consciousness is slow relative to perception. Experiments by Benjamin Libet and colleagues (Libet and Kosslyn 2005, 50–55) show that before subjects indicate that they have decided to raise their arms, the muscle action has already started. Although Daniel Dennett is critical of Libet's experimental design (Dennett 1992), he agrees that consciousness is belated, behind perception by several hundred milliseconds, the so-called "missing half-second." This cost, although negligible in many contexts, assumes new importance when cognitive nonconscious technical devices can operate at temporal regimes inaccessible to humans and exploit the missing half-second to their advantage, as Mark B. N. Hansen points out in *Feed-Forward: On the Future of Twenty-First-Century Media* (2015). The full implications of temporality as it concerns the interplay between human and technical cognition are explored in chapter 6 on automated trading algorithms.

Finally there are the costs, difficult to calculate, of possessing a self aware of itself and tending to make that self the primary actor in every scene. Damasio comments that "consciousness, as currently designed, constrains the world of imagination to be first and foremost about the individual, about an individual organism, about the self in the broad sense of the term" (Damasio 2000, 300). The anthropocentric bias for which humans are notorious would not be possible, at least in the same sense, without consciousness and, even more so, the impression of a reified self that higher consciousness creates. The same faculty that makes us aware of ourselves as selves also partially blinds us to the complexity of the biological, social, and technological systems in which we are embedded, tending to make us think we are the most important actors and that we can control the consequences of our actions and those of other agents. As we are discovering, from climate change to ocean acidification to greenhouse effects, this is far from the case.

NEURAL CORRELATES TO CONSCIOUSNESS AND THE COGNITIVE NONCONSCIOUS

Antonio Damasio and Gerald Edelman, two eminent neurobiologists, have complementary research projects, Damasio working from brain macrostructures on down, Edelman working from brain neurons on up. Together, their research presents a compelling picture of how core consciousness connects with the cognitive nonconscious. Damasio's work has been especially influential in deciphering how body states are represented in human and primate brains through "somatic markers," indicators emerging from chemical concentrations in the blood and electrical signals in neuronal formations (Damasio 2000). In a sense, this is an easier problem to solve than how the brain interacts with the outside world, because body states normally fluctuate within a narrow range of parameters consistent with life; if these are exceeded, the organism risks illness or death. The markers, sending information to centers in the brain, help initiate events such as emotions—bodily states corresponding to what the markers indicate—and feelings, mental experiences that signal such sensations as feeling hungry, tired, thirsty, or frightened.

From the parts of the brain registering these markers emerges what Damasio calls the proto-self, "an interconnected and temporarily coherent collection of neural patterns which represent the state of the organism, moment by moment, at multiple levels of the brain" (Damasio 2000, 174.) The proto-self, Damasio emphasizes, instantiates being but not consciousness or knowledge; it corresponds to what I have been calling the cognitive nonconscious. Its actions may properly be called cognitive in my sense because it has an "intention toward," namely the representation of body states. Moreover, it is embedded in highly complex systems that are both adaptive and recursive. When the organism encounters an object, which Damasio refers to as "something-to-be-known," the object "is also mapped within the brain, in the sensory and motor structures activated by the interaction of the organism with the object" (2000, 169). This in turn causes modifications in the maps pertaining to the organism and generates core consciousness, a recursive cycle that can also map the maps in second-order interactions and thereby give rise to extended consciousness. Consciousness in any form arises only, he maintains, "when the object, the organism, and their relation, can be re-represented" (Damasio 2000, 160). Obviously, to be re-represented, they must first have been represented, and this mapping gives rise to and occurs within the proto-self. The proto-self, then, is the level at which somatic markers are assembled into body maps, thus mediating between consciousness and the underlying material processes of neuronal and chemical signals.

This picture of how consciousness arises finds support in the work of Nobel Prize–winner neurologist Gerald M. Edelman and his colleague Giulio Tononi (Edelman and Tononi 2000). Their analysis suggests that a group of neurons can contribute to the contents of consciousness if and only if it forms a distributed functional cluster of neurons interconnected within themselves and with the thalamocortical system, achieving a high degree of interaction within hundreds of milliseconds. Moreover, the neurons within the cluster must be highly differentiated, leading to high values of complexity (Edelman and Tononi 2000, 146).

To provide a context for these conclusions, we may briefly review Edelman's Theory of Neuronal Group Selection (TNGS), which he calls "neural Darwinism" (Edelman 1987). The basic idea is that functional clusters of neurons flourish and grow if they deal effectively with relevant sensory inputs; those less efficient tend to dwindle and die out. In addition to the neural clusters, Edelman (like Damasio) proposes that the brain develops maps, for example, clusters of neurons that map input from the retina. Neural groups are connected between themselves through recursive "reentrant connections" (Edelman 1987, 45–50, esp. 45), flows of information from one cluster to another and back through massively parallel connections. The maps are interconnected by similar flows, and maps and clusters are also connected to each other.

To assess the degree of complexity that a functional neuronal cluster possesses, Edelman and Tononi have developed a tool they call the functional cluster index (CI) (2000, 122–123). This concept allows a precise measure of the relative strength of causal interactions within elements of the cluster compared to their interactions with other neurons active in the brain. A value of CI = 1 means that the neurons in the cluster are as active with other neurons outside the cluster as they are among themselves. Functional clusters contributing to consciousness have values much greater than one, indicating that they are strongly interacting among themselves and only weakly interacting with other neurons active at that time.

From the chaotic storm of firing neurons, the coherence of the clusters mobilize neurons from different parts of the brain to create coherent maps of body states, and these maps coalesce into what Edelman calls "scenes," which in turn coalesce to create what he calls primary consciousness (in Damasio's terms, core consciousness). Edelman's account adds to Damasio's the neuronal mechanisms and dynamics that constitute a proto-self from the underlying neurons and neuronal clusters, as well as the processes by which scenes are built from maps through recursive interactions between an organism's representations of body states and representations of its relations with objects.

It is worth emphasizing the central role that continuous reciprocal causation plays in both Damasio's and Edelman's accounts. More than thirty years ago, Humberto Maturana and Francisco Varela (1980) intuited that recursion was central to cognition, a hypothesis now tested and extended through much-improved imagining technologies, microelectrode studies, and other contemporary research practices.

SIMULATIONS AND RE-REPRESENTATIONS IN CONSCIOUSNESS

Let us now turn to the processes by which re-representation occurs. Recalling Damasio's strong claim that there is no consciousness without re-representation, representation is clearly a major function of the proto-self, site of the cognitive nonconscious and the processes that feed forward information to core and higher consciousness. These representations constitute the ground upon which the re-representations are formed. In his theory of "grounded cognition," Lawrence W. Barsalou in an influential article gives a compelling account of how re-representation occurs in what he calls "simulation," "the re-enactment of perceptual, motor and introspective states acquired during experience with the world, body, and mind" (Barsalou 2008, 618).

In particular, sensory experiences are simulated when concepts relevant to those experiences are processed and understood by the modes of awareness, in particular the communicating unconscious. He marshals a host of experimental evidence indicating that such mental reenactments are integral parts of cognitive processing, including even thoughts pertaining to highly abstract concepts formulated by higher consciousness. The theory of grounded cognition "reflects the assumption that cognition is typically grounded in multiple ways, including simulations, situated action, and, on occasion, bodily states" (Barsalou 2008, 619). For example, perceiving a cup handle "triggers simulations of both grasping and functional actions," as indicated by fMRI scans (functional magnetic resonance images). The simulation mechanism is also activated when the subject sees someone else perform an action; "accurately judging the weight of an object lifted by another agent requires simulating the lifting action in one's own motor and somatosensory systems" (Barsalou 2008, 624). In order for a pianist to identify auditory recordings of his own playing, he must "simulate the motor actions underlying it" (2008, 624).

The discovery of mirror neurons extends the idea of simulation to social interactions, including the ability to grasp the intentions of another. Barsalou notes that "mirror neurons . . . respond strongest to the goal of the action, not to the action itself. Thus, mirror circuits help perceivers infer an actor's intentions, not simply recognize the action performed" (623). V. S. Ramachandran in *The Tell-Tale Brain: A Neuroscientist's Quest for What Makes Us Human* (2011) emphasizes the role of mirror neurons in empathy as well as in interpreting intentions: "It's as if anytime you want to make a judgment about someone else's movements, you have to run a virtual-reality simulation of the corresponding movement in your own brain" (123).

Perhaps most surprisingly, such simulations are also necessary to grasp abstract concepts, indicating that the thinking associated with higher consciousness is deeply entwined with the recall and reenactment of bodily states and actions. The importance of simulations in higher-level thinking shows that biological systems have evolved mechanisms to re-represent perceptual and bodily states, not only to make them accessible to the modes of awareness but also to support and ground thoughts related to them. In the "grounded cognition" view, the brain leverages body states to add emotional and affective "tags" to experiences, storing them in memory and then reactivating them as simulations when similar experiences arise. Thus the brain in this perspective conceptualizes not primarily through the manipulation of abstract symbols (the cognitivist paradigm) but through its embodied and embedded actions in the world, as noted in chapter 1 discussing Núñez and Freeman (1999) and Varela, Thompson, and Rosch (1992). One consequence of this view is the emphasis it places on the brain's ontogenetic capacities, in which its interactions with the environment reconfigure synaptic networks (Hayles 2012; Clark 2008; Hutchins 1996). We can now appreciate the emphasis that Damasio places on re-representation, for simulations serve as essential parts of the communication processes between the proto-self and consciousness, investing even highly abstract thoughts with grounding in somatic states.

THE IMPORTANCE OF INFORMATION PROCESSING
IN THE COGNITIVE NONCONSCIOUS

The empirical work on human nonconscious cognition has largely concentrated on visual masking. This can be done either by flashing stimuli too brief to be consciously seen, or by exposing subjects to visual patterns in which target stimuli are "hidden" by a complex array of distractor symbols. In a critical review of these experiments, Sid Kouider and Stanislas Dehaene (2007) chart the changing responses in cognitive psychology to the idea that subliminal stimuli can nevertheless be processed nonconsciously and affect subsequent conscious perceptions in a variety of ways. Dating back to the nineteenth century, such experiments continued to generate interest into the twentieth century, with a flurry of activity that peaked in the 1960s and 1970s. Subsequent scrutiny revealed, however, that many of these midcentury experiments were methodologically flawed, primarily by not establishing with sufficient rigor that the stimuli in fact were not consciously perceived. This led to improved experimental designs, and by the twenty-first century, the skepticism of the 1980s had shifted toward a consensus that nonconscious cognition does indeed occur and influences behavior on multiple levels, ranging from motor to lexical, orthographic, and even semantic responses. Nevertheless, there remains a spectrum of opinion about how important nonconscious cognition is and how it interacts with consciousness.

The positive end of the scale is articulated in a review essay by Pawel Lewicki, Thomas Hill, and Maria Czyzewska (1992) surveying results in cognitive psychology, cognitive science, and other disciplines regarding the functions and structures of nonconscious cognition. They cite empirical evidence showing that nonconscious cognition, in addition to pattern recognition, also performs sophisticated information processing including drawing inferences, creating meta-algorithms, and establishing aesthetic and social preferences. Recalling the fact that consciousness is much slower than nonconscious processes, they point out that "nonconscious information-acquisition processes are incomparably faster and structurally more sophisticated" (796). Indeed, they postulate that the ability of the nonconscious to acquire information "is a general metatheoretical assumption of almost all contemporary cognitive psychology" (796), because experimentalists generally assume that research subjects will not be able to tell them how they acquired the knowledge that their behavior demonstrates they have learned. The problem is not just articulating tacit knowledge, as one might suppose, but rather that subjects "not only do not know how they do all those things but [they] have never known it" (796). This ignorance indicates a "fundamental lack of access" by consciousness to nonconscious "algorithms and heuristics" (796), even though such nonconscious processes are "necessary for every perception, even simple ones." Nonconscious cognition, in short, is absolutely essential for higher cognitions, contributing to "the very foundations of the human cognitive system" (796).

The evidence supporting these claims comes from a wide variety of experiments showing that patterns are recognized nonconsciously. When research subjects are prompted to perform the same learning tasks consciously, they are either unable to do so or perform much more poorly than when operating with nonconscious cognition. In one experiment, for example, subjects were asked to locate a target character (in one case, the digit 6) amid an array of distractor symbols. Experimenters found that if there is a consistent correlation between a subtle background pattern and the target, subjects are able to perceive that correlation and learn from it, as indicated by increased performance and faster response times (796). Moreover, even though their improved performance depends on this learning, they are unable to detect the same pattern when consciously attempting to do so. In one experiment, college students were offered a $100 reward to find the "hidden" pattern; even though some participants "spent many hours trying to find the clue . . . none of them managed to come up with any ideas even remotely relevant to the real nature of the manipulation" (798). This result, among others, confirms the idea that the cognitive nonconscious is inaccessible to consciousness, and no amount of introspection will make it so.

An amusing variation involved subjects assumed to be "intellectually capable enough to report any potential introspective experience," namely faculty in a psychology department (797). First they nonconsciously learned a "set of encoding algorithms that allowed them to more efficiently" find the target, as measured by faster identification in a search task. Then the covariation pattern (variables correlated with one another) was changed, and as expected, their performance deteriorated. The subjects had been informed that the test was for nonconscious cognition, but even though they tried hard to discern how the experiment worked, "none of them came even close to discovering the real nature of the manipulation" (797). Instead they speculated that their performance degraded because some kind of threatening subliminal stimuli had been flashed on the screen (798).

In their conclusion, the authors ask whether nonconscious cognition may be considered intelligent. As might be anticipated, they respond by saying that depends on how one defines intelligence. If intelligence means "having its own goals . . . and being able to pursue them by triggering particular actions," then the answer is no (800). However, if intelligence "is understood as 'equipped to efficiently process complex information,'" then the answer is yes (801). The conclusion's significance points in multiple directions. In one sense, it highlights the differences between "intelligence" and "cognition." While intelligence is generally considered an *attribute* that can be quantified, measured, and understood as something either present or not, cognition refers to *processes* that instantiate certain dynamics and structural regularities. This implies that cognition is inherently dynamic and subject to constant change, rather than a durable attribute inherent in a being's makeup. This implication, as we will later see, ties in with how one thinks about information and information processing in general.

In another sense, their conclusion highlights the distinction between goal-directed behavior and information processing. This distinction, however, is not entirely clear-cut, for if information processing provides the framework for subsequent interpretations, as has been shown to be the case in self-perpetuating nonconscious algorithms, then conscious behaviors and goals are always already influenced by inferences that nonconscious cognition has performed beyond the ken of consciousness. The full significance of this point will be explored later in this chapter, as well as in chapter 5, devoted to nonconscious cognitions performed by technical devices.

In summary, the evidence from multiple research studies confirms the speed and complexity of the kind of information processing that nonconscious cognition performs. They reveal nonconscious cognition as a powerful means of finding patterns in complex information, drawing inferences based on these patterns, and extrapolating the learned correlations to new information, thus becoming a source for intuition, creativity, aesthetic preferences, and social interactions. Recalling that subjects were entirely unable to consciously identify patterns that they had already learned nonconsciously, we can appreciate the authors' conclusion that nonconscious cognition is "incomparably more able to process formally complex knowledge structures, faster, and 'smarter' overall than our ability to think and identify meanings of stimuli in a consciously controlled manner" (10). The conclusion underscores one of the major points of my overarching argument: the growing awareness that consciousness is not the whole of cognition, and that nonconscious cognition is especially important in environments rich in complex information stimuli.

INTERACTIONS BETWEEN NONCONSCIOUS COGNITION AND CONSCIOUSNESS

I turn now to consider how consciousness interacts with and is influenced by nonconscious cognition. If its workings are inaccessible to introspection, how does this influence take place? Moreover, does influence flow only from the cognitive nonconscious to consciousness, or does consciousness exert influence on nonconscious cognition as well? These questions are taken up in Stanislas Dehaene's "Conscious and Nonconscious Processes: Distinct Forms of Evidence Accumulation?" (2009), which proposes a theoretical framework that coordinates nonconscious learning with consciousness. He identifies nonconscious cognition with "specialized information processors," citing fMRI evidence (from his own experiments as well as those of others) showing that these specialized processors feed forward their results very quickly, within the first 270 milliseconds after nonconscious perception. These quick response processors have been shown to influence perceptions in a variety of ways, for example by priming subjects either to faster response times if the unseen (subliminal) prime is congruent with the consciously perceived target, or impeding recognition if the prime is incongruent. Moreover, the specialized processors are capable of adding up evidence until a dynamic threshold is reached, after which a response takes place (97). Such a mechanism could account for the capability to learn nonconsciously, for learning would have the effect of lowering the dynamic threshold.

Given this evidence, what are the mechanisms that enable nonconscious cognition to influence consciousness, and vice versa? Dehaene proposes a framework that postulates reverberating circuits of neurons that work through a combination of bottom-up and top-down signals. The temporal dimension here is crucial, for if the feedforward information is contextually appropriate to the (conscious) executive control that determines the focus of attention, then neurons with long-range excitatory axons, which are associated with consciousness, begin to "send supportive signals to the sensory areas that first excited them." Let us suppose, for example, that a dog hears a sound in the woods. If his attention is attracted by it, he pricks up his ears, which happens as a result of top-down supportive signals responding to lower-level sensory excitations. This top-down support from higher levels may send "increasingly stronger top-down amplification signals" until a dynamic threshold is crossed. At this point, "activation becomes self-amplifying and increases nonlinearly," as the areas of the brain associated with consciousness can then maintain activation indefinitely independent of the decay of the original signal.

This phenomenon, which Dehaene calls "ignition of the global workspace," corresponds with the entry into consciousness of the feedforward signals from the specialized processors. Moreover, at this point "stimulus information represented within global workspace can be propagated rapidly to many brain systems," as consciousness enrolls diverse systems throughout the body in coordination with its neural activity. The dog, for example, may now identify the sound as coming from a nearby rabbit and rushes off to give chase. A critical point in Dehaene's analysis of how subcortical processors work is the temporal dimension, for subcortical processors *require* this top-down support in order to endure past the half-second mark. Indeed, he defines a subliminal signal as one that "possesses sufficient energy to evoke a feedforward wave of activation in specialized processors," but "has insufficient energy or duration to trigger a large-scale reverberating state in a global network of neurons with long–range axons."

From this analysis, a useful distinction emerges between nonconscious or subliminal processing and the kind of processing associated with what is called the "attentional blink," popularized in Malcolm Gladwell's *Blink: The Power of Thinking Without Thinking* (2005). In the attentional blink, information does not reach consciousness because the global workspace is occupied by information from another processor. A model for the attentional blink proposed by Dehaene, Claire Sergent, and Jean-Pierre Changeux (2003) has succeeded in predicting "a unique property of nonlinear transition from nonconscious processing to subjective perception. This all-or-none dynamics of conscious perception was verified behaviorally in human subjects" (2003, 8520). In the case of the gorilla walking unnoticed across the basketball court, for example, the model would predict that the global workspace, occupied by the instruction to count the number of passes, will not accept the information being forwarded by nonconscious cognitive processes. Sid Kouider and Dehaene (2007) call this phenomenon of the attentional blink "preconscious processing," defining it as occurring "when processing is limited by top-down access rather than bottom-up strength." Nonconscious or subliminal processing, by contrast, may not receive the necessary top-down support to continue activation, and in these cases it does not have the energy or duration to activate the global workspace on its own, whereas preconscious processing can enter the global workspace once space becomes available. If we return to the example of the gorilla, this explains why some subjects did in fact consciously perceive the intrusion. For some reason, their attention (or executive control) was not so focused on the counting task that it preempted the global workspace entirely, so that preconscious processes containing information about the gorilla could enter the workspace and hence become available to consciousness.

Dehaene points out an important fact relevant to the distinction between preconscious processing and nonconscious cognition: although subliminal primes can modulate the response time, *"they almost never induce a full-blown behavior in and of themselves"* (emphasis added, Dehaene 2009, 101). Assigned a task by consciousness, nonconscious cognition can carry it out efficiently and effectively; it can also integrate multiple kinds of signals both from within the body and without, draw inferences from these signals, and arrive at decisions that adjudicate between conflicting or ambiguous information to create feedforward activation that influences a wide variety of behaviors. Nevertheless, consciousness remains necessary, as Dehaene puts it, for an "information representation [that] enters into a deliberation process" and "supports voluntary actions with a sense of ownership" (Dehaene 2009, 102). In this perspective, nonconscious cognition is like a faithful advisor supporting and influencing consciousness but not initiating whole-body action on its own—in other words, acting much more like Joe Biden than Dick Cheney.

To round out this account of the relation between the cognitive nonconscious and consciousness, we may consider their respective evolutionary roles, a topic discussed by Birgitta Dresp-Langley (2012). She argues that "statistical learning, or the implicit learning of statistical regularities in sensory input, is probably the first way through which humans and animals acquire knowledge of physical reality and the structure of continuous sensory environments" (1). She elaborates that "this form of non-conscious learning operates across domains, across time and space, and across species, and it is present at birth when newborns are exposed to and tested with speech stream inputs" (1). Consciousness, by contrast, "kicks in much later in life, involving complex knowledge representations that support conscious thinking and abstract reasoning" (1). On an evolutionary timescale, nonconscious cognition no doubt developed first and then consciousness was built on top of it, with massive cross-connections between them through what Edelman calls reentrant signaling and other mechanisms. Nevertheless, the limited ability of consciousness to process information, both because of its narrow focus and relatively slow dynamics, means that nonconscious cognition continues to play significant roles in discerning patterns in the environment, processing emotional cues in faces and body postures (Tamietto and de Gelder 2010), drawing inferences from complex correlations among variables, and influencing behavioral and affective attitudes and goals.

An important point to remember from this research is that the cognitive nonconscious not only has the function of forwarding information to consciousness but also of *not* forwarding information not relevant to the current situation. Otherwise, the capacity of consciousness to process information would soon be overwhelmed. As Dresp-Langley observes, "non-conscious representation is aimed at reducing the complexity at the level of conscious processing. It enables the brain to select, from all that it has learnt about outer and inner events, only what is needed for producing a meaningful conscious experience" (7). Dresp-Langley summarizes, "A great deal of human decision making in everyday life occurs indeed without individuals being fully conscious of what is going on, or what they are actually doing and why. Also, human decisions and actions based on so-called intuition are quite often timely and pertinent and reflect the astonishing ability of the brain to exploit non-conscious representations for conscious action, effortlessly and effectively" (7). In short, consciousness *requires* nonconscious cognitive processing of information and could not function effectively without it.

NONCONSCIOUS COGNITION AS A HUMANISTIC CONCEPT: THE MCDOWELL-DREYFUS DEBATE

How might the concept of nonconscious cognition be useful to the humanities? As an example, I take the debate between Hubert L. Dreyfus and John McDowell, two eminent philosophers, on the question of how pervasive rationality is in human experience. Their arguments take place within the disciplinary norms of philosophical discourse, and they illustrate the ways in which those norms can impede discussions about nonconscious cognition. The ostensible subject of the debate was the nature of conscious knowledge, and more broadly about the paradigm, dominant for decades in cognitive science, that sees the brain as a computer running programs and processing abstract symbols. The debate's genesis was Dreyfus's 2005 presidential address at the meeting of the American Philosophical Association, when he took on McDowell's claim in *Mind and World* (1996) that intelligibility is pervaded by rational faculties. McDowell answered, and their respective positions are summarized in their essays in *Mind, Reason, and*

Being-in-the-World: The McDowell-Dreyfus Debate (Dreyfus 2013), along with essays by a number of other philosophers commenting on the issues. At stake is whether ordinary human activities are pervaded by rationality, as McDowell seemed to claim, or whether nonrational processes also have important roles to play.

Dreyfus, long a critic of the paradigm that equates the brain with a computer, argues that much of human life proceeds through what he calls "absorbed coping" (22–23 passim) and that these actions are not fundamentally conceptual. As examples, Dreyfus cites a soccer player in the heat of a game, the distance two interlocutors stand from one another while engaged in conversation, the performance of gender roles, and (somewhat problematically, in my view), a chess master playing lightning chess. These practices, he argues, open "a *space of meaning* that governs all forms of cultural coping" (25). He wants to contest what he characterizes as McDowell's position that "all human activity" can be accounted for either as "shaped natural reactions" or "the space of reasons" (26), and he maintains that "absorbed cultural coping" fits into neither of these categories. Moreover, he argues that absorbed coping is pervasive in human life. We can see in his argument parallels between "absorbed coping" and the nonconscious information processors discussed earlier, although such nuances as the distinction between preconscious processing and nonconscious cognition are blurred as they are mixed together in the "absorbed coping" catchall.

Dreyfus uses as examples of absorbed coping a host of everyday activities, such as opening a door to enter a room or using chalk to mark on a blackboard; he also instances performances by chess masters and other highly skilled individuals, such as professional athletes. He argues that chess masters "learn primarily not from analyzing their successes and failure but from the results of hundreds of thousands of actions. And what they learn are not critically justifiable *concepts* but sensitivity to subtler and subtler *similarities and differences of perceptual patterns.* Thus, learning changes, not the master's *mind,* but his *world*" (35). Although he does not mention nonconscious cognition, the reference to "subtler and subtler" patterns recalls the nonconscious learning in the experiments cited earlier.

In his response, McDowell qualifies his claims to clarify what he means by rational experience. "The idea is not that our experiential knowledge is always the result of determining what reason requires us to think about some question," he explains (42). "Normally when experience provides us with knowledge that such and such is the case, we simply find ourselves in possession of the knowledge; we do not get into that position by wondering whether such and such is the case and judging that it is. When I say that the knowledge experience yields to rational subjects is of a kind special to rational subjects, I mean that in such knowledge, capacities of the sort that *can* figure in that kind of intellectual activity are in play" (42). This is a considerably smaller claim than his original idea that rationality pervades everyday life; indeed it verges on a tautology, saying that organisms capable of reason exercise reason.

Remarkably in this debate, neither philosopher makes a clear distinction between conscious and nonconscious thought, although Dreyfus's "absorbed coping" edges toward the idea of a mode of cognition that is not fully conscious. Also remarkable is the confidence of both men that discourse and argument alone are sufficient to settle the dispute, a self-reinforcing circle of belief in McDowell's case that uses rational argument to elevate the importance of reason. Although both mention a wide range of other philosophers, from Aristotle to Heidegger along with many others, neither cites any experimental research that would bear on the questions they debate. Even more dramatically, neither indicates that conscious processing, including rationality, is limited in its speed and range. When Dreyfus assigns habitual and learned cultural patterns to the category of "absorbed coping," he recognizes the role of habit but fails to articulate clearly how nonconscious cognition supports, and is supported by, conscious actions. Although McDowell recognizes that we may not consciously know how we know ("we simply find ourselves in possession of that knowledge"), he seems entirely unaware of the sophisticated pattern recognition and inference capabilities of the cognitive nonconscious, leading to a systematic overestimation of how important rationality is to everyday life among humans (although he does not explicitly say so, he seems to assume that only humans have rational capabilities).

Dreyfus clearly wants to challenge this conclusion for reasons similar to those articulated in his two books outlining "what computers can't do" (essentially, he argues that embodied human actions in contexts create richer horizons of meaning than computers can ever acquire). However, by remaining within the circle of discursive argument and not making use of empirical evidence such as experiments, statistics, and so forth, he is operating within disciplinary norms of philosophical discourse and hence is more or less bound to try to settle this matter through discursive argument alone. Moreover, even within the narrow circle of argument, he does not always choose the most felicitous means of making the case for absorbed coping. His example of a chess master playing lightning chess casts his argument in an untenable either/or position (the chess master *either* uses absorbed coping *or* he uses rationality). McDowell takes advantage of this slip to point out, rightly, that intellectual analysis has to be part of that activity as well. Although this is only implicit in Dreyfus's argument, the fact that the activity is *lightning* chess is important, for the implication is that the chess master will not have *time* to think, an indirect reference to the fact that conscious thinking is much slower than nonconscious cognition. In fact, it is likely that nonconscious cognition sifts through the information and forwards to consciousness only the decision points where reason has to be invoked. This might occur, say, only 5 percent of the time in which all the decisions must be made, but this 5 percent may be what separates the chess master from the accomplished player, and the grand master from the master. Again, Dreyfus's argument here does not fully make clear the kind of coordination between consciousness and nonconscious cognition explicated by empirical research.

In short, Dreyfus's argument is not as powerful as it might be if he had availed himself of a different kind of rhetoric and frame of reference that would show the *majority* of human information processing is not conscious at all, which, as we have seen, is a proposition now widely accepted in cognitive science. Then the debate could turn, not on the question of whether humans are capable of reason (which seems to be how McDowell wants to reposition his argument), but whether reason is central to everyday human action in the world.

Such a clarification would make it possible to talk about the ways in which nonconscious processing, while distinct from consciousness, is in constant communication with it and supports consciousness precisely by *limiting* the amount of information with which consciousness must deal, so that consciousness, with its slower speed and more limited processing power, is not overwhelmed. The point is not that humans are not capable of reason (obviously an easy fallacy to refute), but that reason is supported by and in fact *requires* nonconscious cognition in order to be free to work on the kinds of problems it is well designed to solve.

McDowell's claim that intellectual activity is central to human life could then be understood not as a *quantitative* claim (how much of human cognition can be counted as rational, on which grounds he is simply wrong about its pervasiveness in all cognitive activity), but rather as a *value judgment* about how much reason is able to accomplish, and how important these accomplishments are to human sociality and modern human life. That nonconscious cognition is absolutely crucial to normal human functioning in the world can then be understood not as a refutation of reason's usefulness, but as the ground for how and why consciousness—and along with it, reason—could develop at all.

In summary, this famous debate, considered sufficiently important to warrant the publication of an essay collection devoted to it and attracting contributions by several other philosophers, misfires because it takes no account of nonconscious cognition, either by McDowell, who would surely be antagonistic to the idea, or by Dreyfus, who moves in a direction that research on nonconscious cognition would support and to which it offers important clarifications. The debate opens a window on how a field such as philosophy, with its emphasis on rational argument, could be both challenged and energized by including nonconscious cognition in its purview. The same could be said of many different disciplines in the humanities, especially in this era when the digital humanities are bringing nonconscious cognition in the form of computer algorithms into the heart of humanistic inquiry, a topic addressed in chapter 8.

PARALLELS TO NONCONSCIOUS
COGNITIONS IN OTHER TRADITIONS

Whereas the previous section discussed how disciplines dedicated to rational discourse can benefit from recognizing nonconscious cognitive processes, this section discusses non-Western and alternative traditions that partially overlap with, and also diverge from, this expanded view of cognition. Mindfulness, my first example, emerged as a meditative practice and stress-reduction technique when it was introduced into Western clinical practice by Jon Kabat-Zinn, among others, in 1979 at the University of Massachusetts Medical School (n.d, "Mindfulness-Based Stress Reduction [MBS]"). It has subsequently been used to treat a variety of disorders, including PTSD, depression, anxiety, and drug addiction, and has achieved wide acceptance within the psychological clinical community and beyond. The technique begins with adopting a correct posture (sitting cross-legged or with a straight back on a chair) and focusing on one's breathing. The beginning meditator will notice almost immediately that her mind begins to wander (a phenomenon explored at length in Michael C. Corballis's *The Wandering Mind: What the Brain Does When You're Not Looking* [2015]); she is advised to note this phenomenon nonjudgmentally, with an alert curiosity and acceptance, returning her focus to her breathing "when it feels right."[1] With practice, the meditator can remain focused for longer periods beyond the initial suggested time of ten minutes, and with more acute awareness of bodily rhythms, the evanescence of conscious experience, and a reflective acceptance of her responses to incoming stimuli, without becoming caught up in the responses themselves. Although Kabat-Zinn argued that the practice is not identified solely with Buddhism and other Eastern religions, citing American transcendentalists Henry Thoreau and Ralph Waldo Emerson, these figures were of course themselves very influenced by non-Western meditative practices.

In terms of the framework developed here, meditative practices have the effect of diminishing investment in the narratives of consciousness, partially or wholly clearing the global workspace, and therefore making room for feedforward information coming from nonconscious processes, particularly about body processes, emotional responses, and present awareness, thus centering the subject and putting her more closely in touch with what is actually happening from moment to moment within her body as it is embedded within the environment. Meditation in this sense is compatible with the framework developed here, adding an experiential component to the intellectual arguments presented thus far.

In the Buddhist tradition, however, clear divergences begin to appear in the epistemological and especially the ontological consequences of meditative practices. The seminal text exploring both the convergences and divergences between nonconscious cognition and meditation is *The Embodied Mind: Cognitive Science and Embodied Experience*, by Francisco J. Varela, Evan Thompson, and Eleanor Rosch (1992). Published more than two decades ago, this text remains an important work articulating embodied/embedded views of cognition together with meditative practices. Focusing specifically on such practices as immersing oneself in the rhythmic activity of breathing (rather than being immersed in the narratives of consciousness), they note that in such moments, one becomes aware that "at each moment of experience there was a different experience as well as a different object of experience" (69). The obvious conclusion is that there is no self as such, only the transient flow of experience. Because consciousness fears that the loss of selfhood would equal death, it tends to panic and to grasp after the illusion of a self. The purpose of meditation is to overcome this reaction, realize the absence of a self (emptiness or sunyata) as an opening out into the world instead of a loss, and begin to explore the experience of awareness within this opening out.

The divergences spring from the further realizations and consequences of this emptiness. In the Buddhist traditions of Madhyamika schools, which seek a "middle way" between objectivism and subjectivism (Varela, Thompson, and Rosch, 229ff.), emptiness extends both to subject and object, self and world. Just as there is no ground for the self, there is no ground for the world either. Neither rests on a transcendental or permanently objective basis. Instead, they bring each other into existence through their interplay, which the authors designate as "codependent arising" (110). As the authors state, this marks a significant difference even with Western traditions that similarly endorse the view that the self is an illusion. "There is no methodological basis [in Western disciplines] for a middle way between objectivism and subjectivism (both forms of absolutism). In cognitive science and in experimental psychology, the fragmentation of the self occurs because the field is trying to be scientifically objective. Precisely because the self is taken as an object, like any other external object in the world, as an object of scientific scrutiny, precisely for that reason—it disappears from view. That is, the very foundation for challenging the subjective leaves intact the objective as a foundation" (230). The upshot, for them, is that they endorse the Buddhist views philosophically but maintain that on a practical, everyday basis, it makes little difference—except, that is, in the case of theories of cognition, where it leads to a framework they call "enactive cognitive science," based on the premise that self is embedded in world and world in self.

Even so, this theory remains only a theory. "Enactive cognitive science and, in a certain sense, contemporary Western pragmatism require that we confront the lack of ultimate foundations. Both, while challenging theoretical foundations, wish to affirm the everyday lived world. Enactive cognitive science and pragmatism, however, are both theoretical; neither offers insight into how we are to live in a world without foundations" (234). Hence the importance for them of Buddhist traditions, where "the intimation of egolessness is a great blessing; it opens up the lived world as path, as the locus for realization" (234). Although it is beyond the scope of the present study to delve more deeply into the ontological questions Varela, Thompson, and Rosch raise, I note that nonconscious cognitive processes do not require and, to a certain extent, actively counteract the need for transcendental certitudes. To this extent, then, the framework developed here is compatible with enactive cognition, although questions about the interplay between objectivism and subjectivism may indeed mark a rupture where divergences appear between my approach and that of Varela, Thompson and Rosch.

The last parallel I want to explore is with speculative realism as articulated in the work of Graham Harman (2011) and Ian Bogost (2012). Similarities include a desire to decenter the human, an interest in inquiring into the modes through which nonhuman others encounter the world, a realization that we can never be bats (as Thomas Nagel pointed out in his famous essay), and the felt need to create an integrated framework through which these ideas might coalesce. As I have written elsewhere (Hayles 2014b), I have several points at which I diverge from their views, including what I see as the dearth of relationality in Harman's model, as well as his assertion (and Bogost's) that objects recede from us infinitely and so can never be known at all, which seems to me obviously contradicted by empirical knowledge in all its forms including science, engineering, medicine, anthropology, and digital humanities. Beyond these particulars, there is also the emphasis on cognition, conscious and nonconscious, that is central to my framework but more or less irrelevant to theirs. Nevertheless, Bogost's exposition of the Foveon-equipped Sigma DP digital image sensor (Bogost 2012, 65–66) can serve as a fine demonstration of how technical cognition works, although this is not how he frames it (he positions it as a caricature or metaphor to make it align with his assertion that we can never know how objects actually experience the world).

To conclude this chapter, I want to zoom out from its detailed expositions and sketch in broad terms my vision of how a member of the Homo sapiens species encounters the world. Alert and responsive, she is capable of using reason and abstraction but is not trapped wholly within them; embedded in her environment, she is aware that she processes information from many sources, including internal body systems and emotional and affectual nonconscious processes. She is open to and curious about the interpretive capacities of nonhuman others, including biological life-forms and technical systems; she respects and interacts with material forces, recognizing them as the foundations from which life springs; most of all, she wants to use her capabilities, conscious and nonconscious, to preserve, enhance, and evolve the planetary cognitive ecology as it continues to transform, grow, and flourish. That, for me, is the larger picture toward which the details documented in chapter 2 point.

CHAPTER 3

The Cognitive Nonconscious and the New Materialisms

Among the promising developments for reassessing the traditional humanist subject are the new materialisms. Their diversity notwithstanding, the theoretical frameworks proceeding under this banner generally argue for a similar set of propositions. Chief among these is decentering the human subject, along with the characteristics that have long been identified with human exceptionalism, including language, rationality, and higher consciousness. Also prominent is the idea that matter, rather than being passive and inert, is "lively" and "vibrant" (Bennett). In some versions of the new materialisms, a strong emphasis on ontology emerges (Barad, Parisi, Braidotti), accompanied by a reframing of ontological premises, often along Deleuzian lines emphasizing metastabilities, dynamic processes, and assemblages (Grosz; Parikka; Bennett). In general, these approaches tend to locate the human on a continuum with nonhuman life and material processes rather than as a privileged special category (Braidotti; Grosz). Finally, they emphasize transformative potentials, often linking these with the capacity for new kinds of political actions (Grosz; Braidotti). After the baroque intricacies of the linguistic turn, these approaches arrive like bursts of oxygen to a fatigued brain. Focusing on the grittiness of actual material processes, they introduce materiality, along with its complex interactions, into humanities discourses that for too long and too often have been oblivious to the fact that all higher consciousness and linguistic acts, no matter how sophisticated and abstract, must in the first instance emerge from underlying material processes.[1]

Despite their considerable promises, the new materialisms also have significant limitations. Conspicuously absent from their considerations are consciousness and cognition, presumably because of the concern that if they were introduced, it would be all too easy to slip into received ideas and lose the radical edge that the focus on materiality provides. This leads to a performative contradiction: only beings with higher consciousness can read and understand these arguments, yet few if any new materialists acknowledge the functions that cognition enables for living and nonliving entities. Reading them, one looks in vain for recognition of cognitive processes, although they must necessarily have been involved for these discourses to exist at all.

A new materialist might object that there are already plenty of discourses, historical and contemporary, that play up the roles of consciousness and cognition, and it is not her obligation to rehearse or amend these in order to foreground materiality. Separating materiality from cognition does not, however, strengthen the case for materiality. On the contrary, it weakens it, for it erases the critical role played by materiality in creating the structures and organizations from which consciousness and cognition emerge. While this is by no means all that the "liveliness" of materiality can do, it is a particularly fraught and consequential form of material agency, and to ignore it leads to a very partial and incomplete picture. Moreover, such erasures encourage overly general analyses, in which crucial distinctions between kinds of material agency are not acknowledged, presumably because to include them would compromise the decentering project. To reason so confuses decentering the human with its total erasure, an unrealistic and ultimately self-defeating enterprise, considering that the success of the decentering project depends precisely on persuading humans of its efficacy.

The framework provided by an expanded idea of cognition can help to offset these limitations (Hayles 2014a). Traditionally, cognition has been identified with human consciousness and thought. As we have seen, this view is now under pressure from the emergence of cognitive biology, a scientific field that advances a much more capacious view of cognition, maintaining that all life-forms have some cognitive capacity, even plants and microorganisms. In the area of human cognition, nonconscious cognition has been shown to perform functions essential to consciousness; moreover, there is growing evidence, as documented in chapters 1 and 2, that most human behavior is not conscious but rather stems from both unconscious scanning and nonconscious processes.

The emphasis on nonconscious cognition participates in the central thrust of decentering the human, both because it recognizes another agent in addition to consciousness/unconsciousness in cognitive processes, and because it provides a bridge between human, animal, and technical cognitions, locating them on a continuum rather than understanding them as qualitatively different capacities. In addition, nonconscious cognition encourages us to recognize distinctions between different kinds of material processes and correspondingly different kinds of agencies. In particular, it distinguishes between material forces that can adequately be treated through deterministic methods, forces that are nonlinear and far from equilibrium and hence unpredictable in their evolution, the subset of these that are recursively structured in such a way that life can emerge, and the yet smaller set of processes that lead to and directly support cognition. Agencies exist all along this continuum, but the capacities and potentials of those agencies are not all the same and should not be treated as if they were interchangeable and equivalent. Finally, the nonconscious cognitive framework provides a countervailing narrative to the Deleuzian concepts and vocabularies pervasive in the new materialisms, recognizing that forces, intensities, assemblages, and the rest are balanced in living systems with forces of cohesion, survival, and evolution. Without such correctives, the enthusiasm for all concepts Deleuzian threatens to ensnare some of the more extreme instances of new materialism in a self-enclosed discourse that, although it makes sense in its own terms, fails to connect convincingly with other knowledge practices and veers toward the ideological, in which practices are endorsed for their agreement with the Deleuzian view rather than because they adequately represent acts, practices, and events in the real world.

Making this case requires careful consideration of the differences between various camps among the new materialisms, along with a rigorous exploration of where and how a nonconscious cognitive framework adds constructively to new materialist projects, where it differs from new materialist claims and provides useful correctives, and where it breaks new ground not considered by the new materialists. To facilitate the analysis, the argument will proceed according to concepts central to new materialisms, including ontology, evolution, survival, force, and transformation.

ONTOLOGY

An outlier among new materialists, Karen Barad derives her brand of materialism from the physics-philosophy of quantum mechanicist Niels Bohr. Faced with experimental evidence for the wave-particle duality in the 1920s, Bohr developed an interpretation distinctly different from that of Werner Heisenberg. As is well known, Heisenberg argued for the "interference" interpretation (the observer interferes with the experiment, which leads to the Uncertainty Principle: the uncertainty in the momentum times the position cannot be less than a minute quantity calculated from Planck's constant). Bohr, by contrast, thought that the issue was more complex. He pointed out that to perform a measurement, the experimenter has to decide on an experimental apparatus. Using a simplified setup for clarity, Barad shows that the apparatus used to measure position is mutually exclusive from one measuring momentum, so the experimenter must choose between them. Consequently, as the apparatus measuring position achieves more precision, the measurement of momentum becomes correspondingly more uncertain, and vice versa. Bohr understood this situation as implying, not that the experimenter has "disturbed" the measurement, but rather that the position and momentum *do not have determinate values* until they are measured. As Barad points out, later experiments confirmed his intuition.

For Bohr, this phenomenon remained in the realm of epistemology. The point for him was that interactions, which include the experimental apparatus and the experimenter, form an inextricable unit determining how reality manifests itself and placing theoretical limits on what can be known. Making the leap into ontology, Barad's strong contribution extends Bohr's insight. She calls the measuring/measurer unit a "phenomenon," explaining that "phenomena are specific material performances of the world" (2007, 335) and coining the term "intraaction" to designate them, "intra" emphasizing that at least two agents must be involved, each bringing the other into existence simultaneously through their intraactions. Thus she answers one of philosophy's first questions: why is there something rather than nothing? A universe without intraactions would in her view be a contradiction in terms, because without intraactions, the universe could not exist as such.

In a course I co-taught with particle physicist Mark Kruse at Duke University, "Science Fiction, Science Fact," focusing on quantum mechanics in science and fiction, we and the students worked through Barad's book together. Mark is part of the team that recently discovered the Higgs boson, and I was interested in his reaction to Barad's claims; as a scientist, he must necessarily believe that the experiments he and his colleagues conduct at CERN indicate something about reality. Of all the scientific fields, particle physics (along with cosmology and cosmochemistry) comes closest to probing experimentally philosophy's first question, although I doubt that anyone in the field would claim he has definitive answers. (When confronting this kind of issue, Mark was fond of saying, "That's a philosophical question," meaning that the question is not susceptible to experimental testing). Nevertheless, particle physics now offers a scenario of the universe's first nanoseconds after the Big Bang (a temporal regime called "inflation"), and it has postulated the mechanisms and curtailments of that stupendous event. No doubt with that background in mind, Mark commented that he thought Barad's vision was both reasonable and consistent with empirical results of his field.

Barad, of course, does not terminate her analysis with quantum mechanics, extrapolating her notion of "agential realism" into discourses, cultural politics, and feminist theory to emphasize the crucial role of inter/intraactions in those fields. Nevertheless, her careful explications of quantum mechanical theories and experiments provide the critical grounding for her project and lend it a certain cachet. She makes the point (not generally recognized) that quantum mechanics applies to macroscopic as well as microscopic objects and that it is our most encompassing, most successful scientific theory to date.[2] Her impressive expertise with quantum mechanics notwithstanding, a skeptical reader may well ask what differences are entailed when her analyses move from elementary particles to organisms, humans, and cultures. Even if the fundamental level of reality is intraactional, does that necessarily imply that cultures are?

Here is where the framework of nonconscious cognition can contribute significantly, for the issue of levels is crucial to it. The specific dynamics operating at different levels provide a way to distinguish between material processes and nonconscious cognition as an emergent result, as well as elucidating the modes of organization characteristic of consciousness/unconsciousness. The framework thus helps to bridge the gap between quantum effects and cultural dynamics, filling in some of the connective tissue that Barad's argument assumes must exist but that she does not explicitly discuss. In this respect she is not unlike most new materialists, for the issue of level-specific dynamics gets short shrift in their discourses, as does the empirical fact that these levels are characterized by different modes of organization. By making clear how some of these distinctions work, the framework of nonconscious cognition offers a useful corrective to new materialist theories and claims. For example, as we saw in chapter 2, nonconscious cognitive processes cannot persist beyond about 500 ms without reinforcement from neurons with long axons involved in the production of consciousness (Kouider and Dehaene [2007]; Dehaene [2009]). Once this top-down reinforcement occurs, what follows is the "ignition of the global workspace," as Stanislas Dehaene calls it, whereby reverberating circuits are activated and thoughts can persist indefinitely. The combination of bottom-up signals with top-down reinforcement illustrates how important distinct levels are in neuronal processes in biological organisms. Similarly scale-dependent phenomena are also found in technical cognitions of computational media, where bottom-up and top-down communications take place more or less continuously.

To understand why new materialisms tend to gloss over levels, we may refer to the philosophy of Gilles Deleuze, which has been enormously influential in the new materialisms. As Elizabeth Grosz observes in her explication of Deleuze, "Deleuze is primarily an ontologist, whose interest is in redynamizing our conception of the real" (Grosz 2011, 55). Writing against the subject, the organism, and the sign (Deleuze and Guattari 1987), Deleuze in the writings he coauthored with Guattari and in his single-author works aims to create a vision that does not depend on those entities and embraces a vitality driven by affects, intensities, assemblages, and lines of flight. He and Guattari acknowledge, of course, that subjects exist, but they highlight the forces that cut transversally across levels and thus do an end-run around most of the concepts populating traditional philosophy. "One side of a machinic assemblage faces the strata, which doubtless make it a kind of organism, or signifying totality, or determination attributable to a subject; it also has a side facing a *body without organs*, which is continually dismantling the organism, causing asignifying particles or pure intensities to circulate, and attributing to itself subjects that it leaves with nothing more than a name as the trace of an intensity" (Deleuze and Guattari 1987, 4).

In the more extreme interpretations of Deleuze, some new materialists focus almost entirely on the "side facing *a body without organs*," eradicating from their narratives the necessary other side to the story, the forces of cohesion, encapsulation, and level-specific dynamics characteristic of living beings, for example in Jussi Parikka's characterization of insects as "mechinological becomings" (2010, 129). As we will see in the following section, this leads either to contradictions, very partial accounts, or significant distortions of scientific practices, especially evolutionary biology. This extreme approach also makes nonconscious cognition, along with the modes of awareness, almost impossible to imagine and certainly impossible to formulate as a formative force in contemporary culture.

EVOLUTION

One of the bolder attempts to apply Deleuzian principles is Luciana Parisi's *Abstract Sex: Philosophy, Bio-Technology and the Mutations of Desire* (2004). This project is noteworthy because, unlike most new materialisms, it creates a framework that recognizes and connects different levels of analyses, including evolutionary biology (the biophysical), sexual reproduction (the biocultural) and biotechnology (the biodigital). Intending to construct a counternarrative to what she calls "Darwinism and neo-Darwinism," Parisi focuses on Lynn Margulis's theory of endosymbiosis, the process by which cells absorbed other freely living organisms in mutations estimated to have occurred 1.5 billion years ago. In this theory, eukaryotic cells (cells with a nucleus and organelles enclosed by a membrane) originated from communities of interacting entities. The idea interests Parisi because she sees it as contesting a view in which natural selection works through heredity, thus privileging the repetition of the same. According to her, in the Darwinian paradigm heredity "crucially designates the economy of self-propagation of the genetic unit (the cause of all differences). Heredity confirms the autonomy of genes from the environment . . . The environment within which the organism is born cannot re-programme the hereditary function of genetic material" (49). This passage makes clear why she would see "Darwinism and neo-Darwinism" as the enemy, for inheritance in this view is a fundamentally conservative force, all the more so since natural selection is understood as a competitive contest for reproduction.

To a large extent, the specter she battles is a paper tiger. This view of evolution may have held true for evolutionary biology in the 1940s, but recent work in epigenetics has shown that DNA is not the whole story of how genes are expressed. Crucially important is gene regulation, carried out by hormones and other chemical signals that regulate when and what genes are activated. These regulatory mechanisms, in turn, have been shown to be affected by environmental conditions (López-Maury, Marguerat, and Bähler 2008). Consequently, gene expression does not, as Parisi would have it, exclude "the feedback relations between environment and genes" (49). Moreover, her argument ignores the Baldwin effect, which traces a feedback loop between mutations in species and the ways in which a species modifies its environment to favor the mutation, another means by which the environment is connected to evolutionary developments. In addition, the Darwinian paradigm assumes that every species is attuned to the possibilities and challenges offered by its ecological niche, defined by its relation to other species and the dynamics of the surrounding habitat. It is simply incorrect, then, to assert as she does that "the environment [in neo-Darwinism] is destined to die as an irrelevant, inert and passive context of development" (49).

Why focus on endosymbiosis? The theory appeals partly because of its displacement of "the zoocentrism of the theories of evolution (the priority of *Homo sapiens*)" (62); in addition, it emphasizes assimilation and networking (Margulis and Sagan 1986) rather than a competitive struggle to survive. To position this theory as *opposed* to "Darwinism and neo-Darwinism," however, misses the point that assimilation and networking are themselves evolutionary survival strategies, particularly if one accepts (as Parisi does) that anaerobic bacteria merged with respiring bacteria as a survival strategy when the earth's atmosphere began to change and the oxygen level rose (63).

These flaws notwithstanding, Parisi's vision works well at the level of endosymbiosis and leads to novel views of "abstract sex," by which she means an analysis that "starts from the molecular dynamics of the organization of matter to investigate the connection between genetic engineering and artificial nature, bacterial sex and feminine desire that define the notion of a virtual body-sex" (10). If this seems difficult to understand, she offers this clarification, in a passage that makes clear her Deleuzian orientation. "Primarily sex is an event: the actualization of modes of communication and reproduction of information that unleashes an indeterminate capacity to affect all levels of organization of a body—biological, cultural, economical and technological . . . Far from determining identity, sex is an envelope that folds and unfolds the most indifferent elements, substances, forms and functions of connection and transmission" (11). The vision of sex as a force that cannot be confined to a subject (whether human or animal) and that permeates life (and nonlife) at all levels, down to and including the molecular, is a compelling insight when applied at the level of bacteria and the functions of assimilation, division, nutrition, and reproduction they carry out.

The vision works less well as Parisi moves to her other two levels, the biocultural and biodigital. The problem here is not so much that her analysis is incorrect as that it operates almost entirely within a Deleuzian perspective, making it difficult to connect her comments with other well-developed discourses in biotechnology, digital media, information, and cultural studies. For example, in the section "Organic Capital," she writes, "With industrial capitalism, reproducibility becomes abstracted from the socio-organic strata through a new organization of biophysical forms of content and expression aiming to subject and regulate masses of decoded bodies (substances of content and expression). Industrial capitalism involves a reterritorialization of decodified socio-organic modes of reproduction (nucleic and cytoplasmic) bringing their sparse codes to the rhythms of mechanical reproduction" (103). The objects of her critique here, including genetic engineering, assisted reproductive techniques such as in vitro fertilization, human and nonhuman cloning, and so forth, are all brought within the purview of "industrial capitalism," without much specificity about what techniques are involved, what the problems are, and exactly how her analysis works to solve them, other than by discursively opening preconceived/preexisting entities to the forces of Deleuzian deterritorialization.

It is not surprising that her analysis works best at the level of bacteria and cells, for here there is no consciousness to complicate the struggle for survival. Surrounded by a permeable membrane defining an inner and outer environment, a unicellular organism interacts with its milieu in ways qualitatively different than more complex organisms, including very short timelines for reproduction and relatively rapid rates of mutation. Its potential for transformation is correspondingly greater than more complex organisms, so Deleuzian terms such as sensitivities to "intensities" and of the cell as a mutating "assemblage" indeed capture some of its significant aspects. As organisms become more complex, cellular dynamics are integrated with many other levels and modes of organization, and the countervailing forces to Deleuzian deterritorialization become correspondingly stronger. Consequently, Parisi must fall back almost exclusively on Deleuzian vocabulary and concepts at the biocultural and biodigital levels, creating a kind of self-enclosed discursive bubble unable to create meaningful links with actual practices in the world.

A midlevel example of biolife forms more complex than single cells but still much simpler than mammals are insects, which may therefore pose an interesting case for how far the Deleuzean dynamics of flow, metamorphosis, and deterritorialization can apply in convincing and persuasive ways. Insects are like unicellular organisms in being devoid of consciousness; like cells, they also have relatively short timelines to reproduce and greater frequencies of mutations (the reason, of course, that fruit flies have been favorites of experimenters for decades). Jussi Parikka has applied Deleuzian ideas to insects, including to the interesting case of insect swarms, where nonconscious cognition emerges as the potential for collective action increases through chemical signaling and other nonsemantic modes of communication.

Referring to von Frisch's pioneering work on bee communication, Parikka argues they are

> not representational entities but machinological becomings, to be contextualized in terms of their capabilities of perceiving and grasping the environmental fluctuations as part of their organizational structures . . . where the intelligence of the interaction is not located in any one bee, or even a collective of bees as a stable unit, but in the 'in-between' space of becoming: bees relating to the mattering milieu, which becomes articulated as a continuum to the social behavior of the insect community. This community is not based on representational content, then, but on distributed organization of the society of nonhuman actors" (129).

The denial of representation is striking, especially in light of the so-called "waggle" bee dance, where the orientation of the bee, the energy it puts forth, and the direction of the dance all communicate precise information about food sources. Why does this not constitute representation, in Parikka's view? It seems that the primary reason is to maintain faithfulness to the Deleuzian paradigm, even when the facts indicate otherwise. Evoked instead is the continuum between the bees and their milieu, intensities as forces that precede and displace the individual, and contingent assemblages. While this vocabulary and set of concepts work well to characterize certain aspects of the behaviors of social insects, they underplay the possibilities for nonconscious cognition and representational actions, an erasure that the framework of nonconscious cognition would help to correct.

This raises the important question whether a middle ground may be forged between Deleuzian becomings and cognition, subjectivity, and higher consciousness. On the one hand, a purist may object that such a middle ground is impossible, because the privilege of origin must be located either with forces and intensities, from which everything else derives (the Deleuzian view), or with the individual subject as a pre-existing entity upon which forces operate. In this view, both cannot have priority simultaneously, and the choice of one or the other entails complex chains of consequences that amount to different worldviews. Suppose, however, that we position them not as contraries forcing an either/or choice but as two different perspectives on an integrated whole (as Deleuze and Guattari seem to suggest in identifying two sides to an assemblage), each with its own truths and insights. In this case, an analogy may be drawn between this situation and the particle/wave duality that Barad discusses. In this analogy, the particle, located as a point mass in space, corresponds to an entity, while wave action, propagating in a nonlocalized manner for a temporal duration, resembles an event. If we ask whether entities or events are primary, from Barad's perspective we are asking the wrong question. Rather we should inquire where are the points of intraaction, the dynamic and continuing interplays between material processes and the structured, organized patterns characteristic of consciousness.

Mediating between material processes and modes of awareness, nonconscious cognition provides a crucial site where intraactions connect sensory input from the internal and external environments ("events") with the emergence of the subject ("entities"). In this view, the nonconscious cognitive framework is positioned not as anti- or pro-Deleuzian but as the mediating bridge between the two perspectives. Among the theorists who have attempted to adopt a similar mediating position (although not in terms of nonconscious cognition) are Elizabeth Grosz and Rosi Braidotti. While both strike some kind of balance between forces/intensities and subjects/organisms, each has her specific way of doing so, along with a distinctive rhetoric and mode of reasoning. We may compare their approaches through the crucial issue of survival, and in this context further explicate the role that nonconscious cognition plays.

SURVIVAL

In her elegant discussion of Darwin, Elizabeth Grosz seeks to align him both with Bergson and Deleuze. Specifically, what she sees in Darwin that contributes to her project of decentering the human is his insistence on the continuum of humans with animals and all of life through evolutionary processes, making the differences between humans and nonhuman animals a matter of degree, not an absolute separation. If the human characteristic of language, for example, is already present in other animals to a lesser degree, then a small step leads to Deleuze's view of life as "the ongoing tendency to actualize the virtual, to make tendencies and potentialities real, to explore organs and activities so as to facilitate and maximize the actions they make possible" (Grosz 2011, 20). Of course, something has to be erased from Darwin to bring about this rapprochement, specifically his assumption that natural selection works on and through the organism (as well as groups of individuals), whereas for Deleuze the organism is what emerges sometimes in metastabilities subject to constant dynamic rearrangements.

In part Grosz sidesteps the organism as the focal point for natural selection through her emphasis on sexual selection. Claiming that sexual selection cannot be reduced to natural selection (as some evolutionary biologists seek to do), she aligns sexual selection with the "force of bodily intensification, its capacity to arouse pleasure or 'desire,' its capacity to generate sensation" (118). This is tied in with an emphasis she sees in Darwin of "the nonadaptive, nonreductive, nonstrategic investment of (most) forms of life in sexual difference and thus sexual selection" (119). Of course, a skeptic can point out that sexual selection is tied to natural selection through competition for mates and thus ultimately for reproductive success, a point Grosz acknowledges but insists cannot be the whole story (120). Indeed, if we think of the peacock's tail and other such extravagances, it is difficult to see what fitness this might signal. Rather, the reasons that peahens prefer one peacock instead of another seem to have very little to do with reproductive fitness and a great deal to do with pleasure, desire, and sensation, just as Grosz argues.

What can the framework of nonconscious cognition add? The implicit conflict between the Darwinian organism and the Deleuzian flow becomes most apparent in Grosz's essay "A Politics of Imperceptibility" (2002). Arguing against identity politics as a feminist strategy, she points out that even an identity acknowledged to be heterogeneous and fractured still assumes that identity will be replicated over time, thus leading to a repetition of the same. For real change to be possible, Grosz argues, one needs a different theoretical orientation, one that emphasizes transformations and openness to constant changes and deterritorializations—namely, the Deleuzian paradigm. Still, it is difficult to see how political agency can be mobilized without some references to subjects, organisms, and signs, the entities that Deleuze writes against.[3]

Nonconscious cognition provides a means by which agency can be located in material processes and in nonconscious cognition as their emergent result, without implying the allegedly stultifying effects of a consciousness unable to transform in relation to its environment.[4] "It is a useful fiction to imagine that we as subjects are masters or agents of these very forces that constitute us as subjects, but misleading," Grosz writes (2002, 471). Nonconscious cognition is the link connecting material forces to us as subjects, thus serving to deconstruct the illusion of subjects as "masters . . . of the very forces that constitute us," without requiring that subjects be altogether erased or ignored as agents capable of political actions.

Notwithstanding the sometimes strained quality of Grosz's argument here, imperceptibility has been given a new purchase on the political by the recent revelations of spying by the NSA and the associated tracking and data collecting of social media, search engines, and the like. Many are now taking down their Facebook pages and trying to erase their presences on the web, so imperceptibility has come to seem a desirable position to occupy. In 2006 Rosi Braidotti anticipated this trend in extending Grosz's argument (while also modifying it) in her focus on "The Ethics of Becoming Imperceptible." Unlike many new materialists, Braidotti acknowledges subjects, with an emphasis on the "sustainable subject" (135). "This subject is physiologically embedded in the corporeal materiality of the self, but the enfleshed intensive or nomadic subject is an in-between: a folding-in of external influences and a simultaneous unfolding-out of affects" (135). Obviously influenced by Deleuze and Guattari, Braidotti in *The Posthuman* nevertheless declares, "I am very independent in relation to them" (2013, 66).

Her independence can be seen in how she conceives the sustainable subject: "Sustainability is about how much of it a subject can take and ethics is accordingly redefined as the geometry of how much bodies are capable of" (136). By "how much of it a subject can take," she means how much a subject can open itself to the "forces, or flows, intensities and passions that solidify—in space—and consolidate—in time—within the singular configuration commonly known as an 'individual' (or rather: di-vidual) self" (136). Her balancing act, then, is to conceive of the subject as an entity open to events, up to a threshold that marks where the subject would disintegrate altogether: "our bodies will thus tell us if and when we have reached a threshold or a limit" (137). Balancing on this threshold, the subject in her discourse sometimes sounds like a stable entity, and at other times, like a momentary assemblage about to disintegrate (note the hesitation above between the "individual" and the Deleuzian "dividual"). The same balancing act is evident in this passage: the subject is an "intensive and dynamic entity . . . [that is] a portion of forces that is stable enough—spatiotemporally speaking—to sustain and to undergo constant fluxes of transformation" (136). Ethics, then, "consists in re-working the pain into threshold of sustainability" (139), a determination to take the coherent self as far into the flux as possible while still maintain its integrity as a self: "cracking, but holding it, still" (139).

By calling this the "sustainable" subject, Braidotti of course implies that survival is a paramount concern. Yet we know that all "individuals" (and even "dividuals") must die. The apparent conflict between death and sustainability is negotiated in her discourse by arguing that death applies only to the individual. Acknowledging that "self-preservation is a commonly shared concern" (146) and that "self-preservation of the self is such a strong drive that destruction can only come from the outside" (146), she nevertheless urges that we see death as "the extreme form of my power to become other or something else" (146), for example, the molecules that survive the body's decay, or if not molecules (many proteins do not survive the death of the individual), then the atoms that persist in and as the worms that have their fill.

Surely the only creatures that can reason so, however, are humans; virtually all other life-forms will struggle to live as long as they can, a sign that they are never reconciled to death. Even as the gazelle feels the lioness's claws in her back, she still desperately kicks to get away. In this sense, Braidotti reinstalls human privilege in the face of death. Although she argues that her framework "implies approaching the world through affectivity and not cognition" (139), surely *only* cognition—specifically, the higher consciousness that humans possess—can achieve the rapprochement with death that she recommends. One wonders, then, why she introduces the idea at all. I conjecture she requires it to achieve a resolution to the conflict between her insistence that the subject is sustainable and her commitment to the Deleuzian paradigm. It is, so to speak, the price of sustaining the balance: the individual disappears but reappears in the Deleuzian flows and intensities with which we are nevertheless urged to identify as somehow another form of "us."

For the positions Grosz and Braidotti articulate, nonconscious cognition offers another interpretive option, a site in which subjects can emerge without implying they are immune to flows and intensities (Grosz's concern), and without requiring that the living human subject should balance at the threshold of disintegration without exceeding it (Braidotti's emphasis). Moreover, as shown in chapter 1, there is considerable empirical evidence that the kind of neurological structure giving rise to nonconscious cognition exists throughout the animal kingdom, including but by no means confined to humans.

The issue of whether a discursive or ideological position has empirical support is, of course, complex, since the chains of reasoning involved in arriving at such conclusions are necessarily permeated with numerous assumptions about what constitutes evidence, what standards of confirmation are entailed, etc. Nevertheless, in my view a position that can claim empirical support is preferable to one that cannot; otherwise, as Bruno Latour has pointed out, it is impossible to distinguish between what is actually the case and what is ideologically driven fantasy. Not surprisingly, the Deleuzian paradigm does not place much (if any) emphasis on empirical verification, preferring to talk about "royal sciences." These, according to Deleuze and Guattari, are concerned with the discovery of abstract laws and general principles, in contrast to the "minor sciences," concerned with heterogeneous materials and craft-like approaches to flow and other phenomena difficult to mathematicize (Deleuze and Guattari 1987, 398–413, esp. 413). Nonconscious cognition subverts this distinction, because it is inherently difficult to measure and yet has strong empirical confirmation from a range of experiments (see Lewicki, Hill, and Czyzewska). Bridging the gap between the mainstream "royal" and marginalized "minor," it challenges the belief that most human behavior is directed by consciousness, without requiring that we accept the ideologically laden assumption that the "minor" or marginal is inherently superior to the "royal" or major.

FORCE

"Force" is often invoked in the Deleuzian paradigm, but seldom is this designation made more specific or precise. In this worldview, force is an essential concept, for if subjects are absent, agency must be located somewhere, and "force" becomes a kind of agentless agent, the driving desire that brings about and also participates in "agencement" (usually translated as "assemblage," a noun perpetually at risk of losing the Deleuzian emphasis on the eventful). Grosz, for example, writes of the "play of the multiplicity of active and reactive forces that have no agency, or are all that agency and identity consist in. Which is to say, force needs to be understood in its full sub-human and super-human resonances: as the *inhuman* . . . which both makes the human possible and which at the same time positions the human within a world where force works in spite of and around the human, within and as the human" (2002, 467).

The eloquence of this passage notwithstanding, it remains extremely imprecise about the nature of "force" and fails to distinguish between different kinds of forces, although these kinds of distinction have been extensively investigated in various scientific fields. On the atomic and molecular levels that Parisi invokes, for example, four fundamental forces are recognized: strong, weak, electromagnetic, and gravity. In chemistry, other kinds of forces come into play in solutions and suspensions, leading to the possibility for self-organizing dynamics to come into play for far-from-equilibrium systems. Parisi often invokes this language, but also fails to notice the importance of systems at equilibrium, whose dynamics can be described by linear differential equations. A great deal is known about when a system's behaviors can be accurately predicted by such equations (shooting a rocket to the moon, for example, or braking a car under specified weather conditions).

The privilege that Parisi and others accord to nonlinear dynamics is often associated with unpredictability, and hence implicitly with the alleged inability of science to deal adequately with such systems. This ignores the important new field of simulation science, where chaotic and complex systems are modeled and yield reliable knowledge about how such systems will behave (Parikka is atypical in his interest in computer simulations on insect swarms and other swarming and schooling behaviors). Moreover, the vagueness of "force" elides an issue that ought to be crucial to the new materialisms: the differences between material forces whose actions are deterministic and hence can be calculated precisely as the sum of the relevant forces, and those that involve self-organizing, chaotic, and complex dynamics and whose actions can lead to the emergence of increasingly complex outcomes, including life and cognition. Among this group, there is also a crucial distinction between systems that are adaptive and those that are not. Both the BZ (Belousov-Zhabotinsky) reaction in chemistry and bacteria's endosymbiosis history are examples of self-organizing systems, but the bacteria are adaptive and can change when conditions change, whereas the BZ reaction, although unpredictable in the various visual displays it creates, cannot adapt in the same way.

In attributing agency to nonhuman forces, these kinds of distinctions are critical. A rock thrown against a window, for example, can be said to act as an agent when it breaks the glass; in this case, its trajectory and force are entirely deterministic and can be calculated precisely if the relevant factors are known. A different kind of agency is exhibited by an avalanche, capable of killing humans and other life-forms and releasing energies on an awesome scale. Unlike the thrown rock, it involves criticality thresholds, which means that it may be difficult or impossible to predict exactly when it will take place. Nevertheless, the agency here is not intentional or mysterious; if all of the relevant factors are known, it can be modeled so as to arrive at a reasonable estimate of how it will behave (the same can be said of earthquakes, where models can predict likely sites for earthquakes and a rough estimate of the span during which they are likely to happen, although the models are not good enough to predict exactly when). Yet other cases are the agencies of systems capable of self-organizing dynamics; here truly surprising results may emerge, the preeminent example of which is the emergence of life millions of years ago in the planet's history. All of these may be said to demonstrate the agency of material processes and the importance of nonhuman forces, but such generalizations are vapid without more precision about the kinds of dynamics and structures involved.

Why keep "force" so vague, then, when so much is known about different kinds of forces and the different agencies they embody? As soon as agency is discussed in the terms indicated above, the mysterious effects of "force" driving the Deleuzian paradigm evaporate into a collection of known agencies. Even when different kinds of agencies are acknowledged, there is a tendency to privilege those that lead to complexity and self-organization (evident in Parisi, for example), valuing the nonlinear over the linear, and the far-from-equilibrium over systems at equilibrium, presumably because these are the forces that lead to novel and unexpected results. Yet these same nonlinear systems are the ones from which life emerged. One might logically suspect, then, that embedded in these preferences is an implicit trajectory that would privilege the living over the non-living, the complex and adaptive over the simple and deterministic. This result is forbidden, however, by the overall aim of decentering the human and celebrating the nonliving as fully capable of agency. Without further specification about the different kinds of agencies and forces, this contradiction indicates that the preference for one kind of force over another is an ideological choice, not an empirical conclusion.

The framework of nonconscious cognition differs from the majority of new materialisms by being explicit about structures, dynamics, and organizations (i.e., "forces") at multiple levels across the human, animal, and technological spectrum. Implicit in the framework is an emphasis on cognition in general, and thus a belief that cognition is important and worthy of study. Indeed, in arguing for nonconscious cognition, the framework aims to *increase* the kinds of acts that are seen as cognitive, especially those in which consciousness is not involved. In this sense, then, it can be said to enlarge the realm of the cognitive as a special kind of capability that emerges from, and yet is distinct from, the material processes that underlie it. To explore further what this kind of approach has to offer, I turn now to transformation, another topic crucial to new materialisms.

TRANSFORMATION

Transformation is typically highly valued in new materialist discourses, for a variety of reasons: the hope for constructive changes within the political scene (Grosz, Braidotti); a kinder, more eco-friendly world not centered around humans (Parisi, Shukin, Bennett); and the possibility of opening up productive changes within humans themselves (Braidotti, Parikka). These are important and significant goals, and the idea of locating agency within material processes is an intriguing possibility, especially given the desirability of locating agency other than in human actors. Nevertheless, the largest transformative forces on the planet today are undoubtedly human agency and human interventions, the effects of which are being registered in climate change, the worldwide loss of habitat for nonhuman animals, the idea of the Anthropocene, and in the reality that human actions are unleashing forces far beyond our ability to control them.

It would seem, then, that a discussion of transformation must necessarily involve recognition of human agencies and the recent exponential growth of nonconscious cognition in technical objects. Jane Bennett, arguably less indebted to Deleuze than some other new materialists (although she mentions him and uses some Deleuzian vocabulary), recognizes the interpenetration of technics and humans in her references to Bernard Stiegler. As I have argued elsewhere (Hayles 2012), this interpenetration applies not only to the dawn of the human species in the Pleistocene era but also in the present, especially in the deep technological infrastructure affecting everything from human directional navigation to the neurological structures activated by reading on the web.

Bennett (2010) makes an important point with regard to this interpenetration: namely the implication that human agency is always distributed, not only within the body between consciousness and nonconscious faculties, but also between the body and the environment. Her examples include edible matter (affecting the body through nutrition), minerals (through bone formation, for example), and worms (whose "small agencies" [96] can be seen in the transformation of forest into savannah as recent research has shown, an example not mentioned specifically by Bennett). Allowing herself the speculation that the "typical American diet" may have played a role in "engendering the widespread susceptibility to the propaganda leading up to the invasion of Iraq" (107), she clearly wants to make connections across multiple levels of analysis, but her focus on material processes makes such an idea almost impossible to document or even to explore. Adding nonconscious cognition into the picture, especially in relation to drones, unmanned autonomous vehicles (UAVs), and other technical devices, would help to bridge the chasm that currently yawns between her examples and her speculations.

In conclusion, a robust account of material processes should not be the end point of analysis but rather an essential component of a multilevel approach that ranges from the inorganic to the organic, the nonhuman to the human, the nonconscious with consciousness, and the technical with the biological. While many new materialists might argue that far too much consideration has been given to the "entity" side of these intraactions, resulting in a devaluation of material processes, dwelling entirely on the "event" side fails to capture essential characteristics of the living, especially the ability of living organisms to endure through time, construct as well as interact/intraact with their environments, and deploy agencies that are not merely emergent but also intentional, even when nonconscious. While it is likely that no one approach can do all this, nonconscious cognition can supply essential components presently absent from most new materialist analyses.

Politically, nonconscious cognition grants to a technological object the privilege of what amounts to a worldview, thus linking its behaviors to the nature of the sensors and actuators that together constitute and define its capabilities.[5] While living organisms (with a few exceptions) must be understood retroactively (for example, by reverse engineering the evolutionary processes), technical objects have been *made*. Leaving aside emergent results (a special case that requires careful orchestration to succeed), each technical object has a set of design specifications determining how it will behave. When objects join in networks and interact/intraact with human partners, the potential for surprises and unexpected results increases exponentially.

The Deleuzian paradigm contributes an enhanced appreciation for nonliving technical objects to generate surprises, new potentialities, and mutating assemblages. The nonconscious cognitive framework supplies a nonreductive empirical approach that enlists the cognitive powers of humans, along with a precise analysis of the structures and organizations involved, while also insisting that nonhumans have cognitive powers of their own. This is not exactly new materialism, vibrant materiality, imperceptibility, or nomadic subjectivity, but rather a paradigm that, cognizant of scientific and technical knowledges, nevertheless strives to bring about a transformation of traditional views of the place of the human in the world.

The Costs of Consciousness: Tom McCarthy's *Remainder* and Peter Watts's *Blindsight*

Like the dark harmony running underneath a bright melody praising the virtues of consciousness, the suspicion that consciousness may not be all it is cracked up to be (by consciousness itself) runs through the history of Western thought. In the late nineteenth and early twentieth centuries, movements such as surrealism and practices like automatic writing sought to crack open the conscious surface and let something else—less rational, less dedicated to coherence—emerge. In the late twentieth century, this tendency began to assume sharper edges, honed by neuroscientific research on brain traumas and other neurological anomalies and benefiting from improved diagnostics, particularly PET and fMRI scans. Popular accounts of neurological deficits, such as Oliver Sacks's *The Man Who Mistook His Wife for a Hat* (1998) and Antonio Damasio's *Descartes' Error: Emotion, Reason, and the Human Brain* (1995), called attention to the crucial role of nonconscious processes in supporting normal human behavior and the inability of consciousness, stripped of these resources, to carry on as if nothing had happened.

As these ideas spread through the culture, writers began to pick up the beat, exploring cracks in consciousness through such works as Richard Powers's *The Echo Maker* (2007), Jonathan Lethem's *Motherless Brooklyn* (2000), Mark Haddon's *The Curious Incident of the Dog in the Night-Time* (2004), R. Scott Baker's *Neuropath* (2009), and Ian McEwan's *Enduring Love* (1998). Criticism followed suit, proposing the category "neurofiction," typified by the special issue of *Modern Fiction Studies*, edited by Stephen J. Burn, "Neuroscience and Modern Fiction" (61.2, Summer 2015).

Amid this profusion, two works stand out for the incisiveness with which they analyze the costs of consciousness and explore their cultural, economic, evolutionary, and ethical implications: Tom McCarthy's *Remainder* and Peter Watts's *Blindsight.* Whereas *Remainder* focuses on an unnamed narrator who, as the result of a never-specified accident, has lost the functionalities that nonconscious cognition performs, *Blindsight* widens the canvas, representing anomalous forms of consciousness and exploring the stakes entailed by the evolutionary road that Homo sapiens traveled when the species (and other terrestrial life-forms) attained consciousness. In significant ways, both novels are influenced by current neuroscientific research, and yet they do not merely follow where science leads. Rather, they interrogate the consequences of consciousness far beyond what the science reveals, probing especially its phenomenological and cultural dimensions. Together, they highlight the crucial importance of nonconscious cognition, in *Remainder* through its loss, and in *Blindsight* through an alien species that has developed a technology vastly superior to earth's even though they are entirely lacking in conscious thought. Together, the novels show how extensively assumptions taken for granted in traditional Western cultures are undermined and even negated when the primacy of higher consciousness becomes questionable, including its association with authenticity, its ability to give (human) life meaning, its identification with rational actor economic theory, its entwinement with the development of sophisticated technologies, and the perceived superiority it bestows on humans as the most cognitively advanced species on the planet (and beyond).

REMAINDER: CONSCIOUSNESS VERSUS THE TENACIOUS POWER OF MATTER

Against a background vagueness surrounding the accident in *Remainder,* two details stand out with sharp clarity: the narrator has been damaged neurologically, and in compensation has received a settlement of eight-and-one-half-million pounds. Although the exact nature of his injury is not specified, we learn that he has lost motor control over the right side of his body (hence damage in the left hemisphere), and that he undergoes therapy, "re-routing" (19) his synaptic networks so he can move his limbs again. Although some functionality is restored, it is not the fluidity most of us take for granted.

Oliver Sacks recounts the case of the "Disembodied Lady" (1998, 43–54), a woman who lost her proprioceptive sense and, as a consequence, could move only by focusing consciously on the motion she wanted, as if she were a puppeteer controlling her puppet-body. She describes herself as "pithed," and something similar seems to be the narrator's case. He learns to run a simulation of the desired motion in his consciousness over and over, presumably training his synaptic networks to pick up the slack created by his injury, but when he moves from simulation to reality, the carrot he has lifted many times in his imagination proves to be obdurate—lumpy, hairy, full (as he perceives it) of spiteful agency. "My undoing: matter," he comments (17), an observation that in a narrow sense describes the object hitting him out of the blue in his accident, and in a broader metaphoric sense, the struggle that will overtake and dominate his life.

The form that struggle takes strongly suggests that another casualty of his accident is nonconscious cognition. With it knocked out of the loop, consciousness must try to perform the tasks normally handled by nonconscious cognition, including the detection and extrapolation of patterns, the integration of somatic markers into coherent body representations, and the fusion of diverse temporal and spatial events so they attain simultaneity. Of course, the primary function of nonconscious cognition is to keep consciousness, with its slow uptake and limited information-processing ability, from being overwhelmed, so its absence means that consciousness is always teetering on the brink of information overload.

With connection to body and world rendered tenuous, the narrator's consciousness compensates by seeking more and more control, to the point of obsession. This, combined with the settlement, impels him to embark on his "re-enactments." The reenactments stage a struggle between an unpredictable and constantly transforming world of matter, and the narrator's attempt to wrest it into familiar patterns he can "capture" by reproducing them under conditions he controls. He begins with his own body, practicing mundane gestures such as opening the refrigerator door again and again until his shirt brushes in a certain way against the counter edge, the door opens smoothly but not too easily, and so forth. When he succeeds in getting it "right," he is rewarded with tingling along his spine and other somatic signals that make him feel, for that instant, as if he is an authentic living being. His obsessive repetitions appear to be attempts to create artificially the modal brain simulations that, as Lawrence Barsalou (2008) has argued, are essential to normal human functioning.[1] Unable to do so through internal mental processes because of his injury, he attempts to externalize them, although as he soon discovers, they are poor substitutes for doing it naturally. The feeling they momentarily bestow on him never lasts, and as it fades, he is driven to try to recapture it through laborious repetitions and reenactments that become progressively more bizarre.

Along with kinesthetic fluidity, the narrator seems also to have lost the neurological capacity for empathy, perhaps by damage to mirror neurons.[2] This, coupled with his sudden fortune, allows him to acquire "staff," which he treats as if they were his personal slaves to dispose as he wishes—reenactors, set designers, and most essential of all, Nazrul Ram Vyas from Time Control UK: "Naz facilitated everything for me. Made it happen" (67). The narrator, unable to feel connected to the world through embodied action, attempts to re-create the connections through conscious introspection, as when he first meets Naz and, at his request, Naz makes a cell phone call: "I traced a triangle in my mind from our restaurant table to the satellite in space that would receive the signal, then back down to Time Control's office" (87). When the narrator happens upon workmen laying wires below street level and considers the connections they enable, he pronounces them "more than Brahmins: gods, laying down the wiring of the world, then covering it up—its routes, its joins" (120).

Whereas the narrator shows the consequences of consciousness operating without nonconscious cognition, Naz embodies the cognitivist paradigm of consciousness that calculates by manipulating formal symbols, as if entirely independent from the inputs supplied by embodied actions and modal sensations. When the narrator explains to Naz what he wants, Naz's "eyes went vacant while the thing behind them whirred, processing. I waited until the eyes told me to carry on" (89). Together, Naz and the narrator represent a vision of human cognition that increasingly takes on nightmare proportions: rationality acting without empathy, decisions made without support from nonconscious cognition, actions undertaken without the connections created by embeddedness in the world.

(RE)ENACTING THE DYSFUNCTIONALITIES OF CONSCIOUSNESS

The catalyst for the narrator's first reenactment, a crack on a bathroom wall, illustrates the peculiar combination of chance and fanatical exactitude that characterizes his projects, as if he were channeling John Cage's "chance operations" while gulping amphetamines. His imagination, growing outward from the crack, conjures an entire building, a mishmash of memories, second-hand anecdotes, and fantasized encounters. We know, for example, that the arrangement he dictates for the courtyard comes not from his own experience but from his friend Catherine's account of the place in her childhood where she felt most authentic, "swings, on concrete . . . And there was a podium, a wooden deck, a few feet from the swings' right" (76). The narrator appropriates her account to re-create this exact scene (122).

Auditioning enactors, he chooses each according to the image he has of them, specifying precisely how they should act. When he doesn't have a clear image, for example of the building's concierge, he demands she wear a blank mask, a white hockey protector that hides her face. Instructing the enactor playing the "liver lady" (so-called because he orders her to fry liver all day), he can't decide what she should say as he passes her while she puts a garbage bag in the hall, so "rather than forcing it—or, worse, just making any old phrase up—I'd decided to let her come up with a phrase" (143), as if the actions and words he dictates are somehow other than arbitrary. His sense of operating within tight constraints makes it seem as if some prior reality is laying down the law he follows, but narrative cracks like this (metaphorically recalling the bathroom wall crack) reveal that the determining force is nothing other than the vignettes and narratives springing forth from his consciousness, shot through with fictions though they may be.

Clues to this fictionality are everywhere, for example, in the passage describing his first meeting with Naz: "He looked just like I'd imagined him to look but slightly different, which I'd thought he would in any case" (85). Much later, when Naz interrupts to convey what he considers important information while the narrator is preoccupied by trivia, the narrator shouts, "No!" "You listen, Naz: *I* say what's important." He continues, "I could see him running what I'd just said past his data-checkers, and deciding I was right: I *did* say what was important. Without me, no plans, no Need to Know charts, nothing" (272). Consciousness here seemingly achieves its dream not only of narrating the world but acting as the dictator determining what goes into that narrative in the first place. In this sense, *Remainder* presents an exaggerated vision of what we may call the *imperialism* of higher consciousness, magnifying to nightmarish proportions its tendency to insist that it alone is in control and is the sole originator of human agency.

Antonio Damasio, in a passage cited in chapter 2, points out that consciousness tends to focus on the individual, making it the center of action: "I would say that consciousness . . . constrains the world of the imagination to be first and foremost about the individual . . . about the self in the broad sense of the term" (Damasio 2000, 300). To the extent that focus on the self is associated with the individualism inherent in liberal ideology, capitalism, and environmental predation, consciousness, especially higher consciousness, participates in and helps to solidify the excesses of consumerist culture. In *Remainder,* this process is exaggerated into maniacal obsession. Sailing through the world without the ballast provided by nonconscious cognition, the narrator's consciousness attempts to seize total control of the environment, with the result that his sense of self swells to grotesque proportions. As his obsession grows, the self's desires are treated as if they were absolute law, regardless of the costs to anyone else.

The economics of the process are clear: the narrator simply *buys* the personnel necessary to his schemes, paying handsome salaries plus bonuses to his "staff" and forking over bribes to everyone else. In this sense, he is the ultimate economic rational actor, bending others to his will so that his (psychological) payoff in the Nash equilibrium matrix trumps everyone else's. In return for the money he distributes (which he regards as mere trash and has only a hazy idea of the amounts involved), he demands unquestioning obedience. For example, when the reenactor charged with tinkering with a motorbike in the courtyard inadvertently spills some oil, the narrator tells him to leave the oil mark because "I might want to capture it later" (144). "Capture?" the reenactor asks, whereupon the narrator thinks irately, "It wasn't his business to make me explain what I meant by 'capture.' It meant whatever I wanted it to mean: I was paying him to do what I said. Prick" (144). In another instance, the narrator is shocked to see the pianist creeping down the stairs even as he hears music coming from the pianist's apartment. Stunned, he demands an explanation, whereupon the pianist sheepishly admits he had made a recording to play when he was away on other business. Although the sensory stimulus the narrator receives is the same whether the music is live or recorded, he goes "white with both rage and dizziness" (157) and tells Naz to "give him hell! Really bad! Hurt him!" and then qualifies, "Metaphorically, I mean," ambiguously adding, "I suppose" (159). Still later he remarks, "I wasn't bound by the rules [that he arbitrarily lays down]—everyone else was, but not me" (225).

If capitalism alienates the worker from his labor, as Marx argued, the narrator uses money in a fashion even more alienating than an Industrial Age robber baron, demanding mind-numbing labor from his reenactors for hours on end without a break and without even the satisfaction of producing a tangible product, other than the temporary easing of his insatiable desire for repetitive patterns. "I generally put the building into *on* mode for between six and eight hours each day," he remarks. "Sometimes there'd be a five-hour stretch" (161). Preoccupied with his own repetitive gestures, as when he spends an entire day practicing brushing past his sink, he sometimes forgets to reset the building to *off,* leaving his reenactors to cope as best they can.

TEMPORIZING TEMPORALITY

The satisfaction the narrator receives is fleeting, so he must constantly introduce new tweaks, new realms over which he can gain control. At first this extension is achieved through a model of his building that he commissions, enabling him to position within it toy figures, which he then duplicates in the actual building. For example, the figurine he uses to stand in for the motorbike enthusiast is kneeling, so he requires the motorbike enactor to kneel first on the pavement and then on the swing, making life correspond to the simulation—a significant dynamic that expands to ominous proportions. When he has squeezed all the juice from the model, he goes further afield, drawn by chance to a shop where he gets a car tire fixed. He finds three boys there and asks the oldest where the "real people" are, and the boy responds, "I'm real" (168). Afterward the narrator orders an exact reproduction of the shop built inside an airplane hangar. As he has the tire-fixing scene reenacted, he reports, "I experienced a sensation that was halfway between the gliding one I'd felt when my liver lady had spoken to me on the staircase during the first reenactment in my building and the tingling that had crept up my right side on several occasions. This mixed sensation grew as we reached the part where the boy intoned the words: "I—am—real" (177). It is not a coincidence that his feeling peaks at this line. Much later, the narrator explains that his *only* goal throughout the reenactments is "to allow me to be fluent, natural, to merge with actions and with objects until there was nothing separating us—and nothing separating me from the experience that I was having; no understanding, no learning first and emulating second-hand, no self-reflections, nothing: no detour. I'd gone to these extraordinary lengths in order to be real" (247).

Having lost the fast information-processing capability of nonconscious cognition, the narrator cannot make his consciousness speed up faster than its belatedness allows, so he compensates by ordering the world to slow down. He explains to the pianist that he wants him to "Start out at normal—no, at half speed—and when you slow down, when you're in the most slowed down bit of all, just hold the chord for as long as you can" (224). To the concierge, whom he had ordered to remain static, he says "now I want you to do nothing even slower . . . 'What I mean,' I told her, 'is that you should think more slowly. Not just think more slowly, but relate to everything around you slower. So if you move your eyes inside your mask, then move them slowly and think to yourself: *Now I'm seeing this bit of wall, and still this bit, and now, so slowly, inch by inch, the section next to it, and now an edge of door, but I don't know it's door because I haven't had time to work it out yet*—and think all this really slowly too'" (224). The monomaniacal nature of his obsessions expands so far that when he times the sun's passage across the hall floor and finds it doesn't match his previous observation, he complains to chief staff members Annie and Frank, "The sunlight's not doing it right" (224). At first bewildered, Annie finally grasps his meaning and explains that the time difference is because weeks have passed, and the "sun's at a different angle to us than it was" (229). Although the narrator quickly tries to cover over the gaffe, it reveals the extent to which he believes he can control his environment.

It also implicitly reveals that the time-slowing command only works with his "staff," people whom he has hired to do his bidding. In the world at large he needs another strategy, one he discovers by hearing a coach shout to his players, "Take your time. Slow each second down" (238). "This was good advice," he muses, and begins to practice slowing down his own perceptions, letting the instants stretch out until he can move within them at leisure. As we have seen, one of the functions that nonconscious cognition performs is integrating discrete temporal events (occurring at different points within a window of about 100 milliseconds) into perceived simultaneity. Operating in the absence of this nonconscious mechanism, the narrator's consciousness is free to carry out its own version of time manipulation, slowing down the action at decisive moments as if reality was being screened inside his head as a slow-motion film.

With this strategy in place, the narrator moves in two directions at once—further into reenactments, hiring actors to reenact the actions of his own enactors in endless regression, and into the real world, where simulation and reality start to melt into one another. As the logistics become more complex, his consciousness, already overloaded with information it must try to process without the help of nonconscious cognition, begins to short out altogether as he repeatedly drifts off into "trances," blackouts lasting for hours or days. Naz, scrambling to keep up with the narrator's increasingly ambitious plans, draws more and more on his computational abilities. The narrator comments, "I could almost *hear* the whirring: the whirring of his computations and of all his ancestry, of rows and rows of clerks and scribes and actuaries, their typewriters and ledgers and adding machines, all converging inside his skull into giant systems hungry to execute ever larger commands" (234–35). Faced with the narrator's ultimate scheme—pretending to stage a bank robbery reenactment while actually robbing a bank—Naz becomes, as the narrator observes, "drunk: infected, driving onwards, on towards a kind of ecstasy just by the possibilities of information management my projects were opening up for him, each more complex, more extreme. My executor" (235). "Thank you," Naz replies to the narrator; "I've never managed so much information before" (235).

ADDICTING TRAUMA

As the narrator's consciousness grows increasingly precarious, some portion of his psyche splits off and reappears in ventriloquized form as a short man, a borough councillor whom Naz identifies as such (239) but who assumes an alter hallucinated existence as an objective reporter on the narrator's thoughts and condition. The tip-off that the character's lines are hallucinated comes when he says he smells cordite (238), an odor the narrator notices on several occasions but no one else can detect. Within the diegesis, the hallucination suggests that narrating consciousness recognizes its own dysfunctionality and creates this figment to give accurate accounts of the narrator's phenomenological experiences and the reasons he pursues his increasingly delusional goals. From a perspective outside the diegesis, the character allows the author, as the narrator plunges deeper into dysfunctionality, to offer explanations to the reader. As the distance widens between the narrator's perceptions and the (presumed) reader's commonsense experiences of the world, the author risks the reader's incredulity; the "short councillor" forestalls this possibility by mediating between the two perspectives, suturing one to the other. In a similar vein are Doctor Trevellian's comments. Summoned by Naz when the narrator first falls into a trance, Trevellian explains that the brain, faced with trauma, manufactures its own endogenous opioids, in effect making the traumatized person into a home-grown addict, so that he returns again and again to the traumatic source to get another fix (220).

REAL SIMULATIONS

What these explanations cannot completely mask is the growing insanity of the narrator's actions. Reading about a bank robbery, he first orders a reenactment of it but then makes "a leap of genius": "a leap to another level, one that contained and swallowed all the levels I'd been operating on up to now . . . lifting the re-enactment out of its demarcated zone and slotting it back into the world, into an actual bank whose staff didn't know it as a re-enactment: that would return my motions and my gestures to ground zero and hour zero, to the point at which the re-enactment merged with the event. It would let me penetrate and live inside the core, be seamless, perfect, real" (265). The plan generates oxymorons that Baudrillard (1995) would have loved: a real reenactment, or a reenacted real. Naz immediately spots the logistical problem (although stunningly blind to the ethical one): preventing the enacting bank robbers from realizing that the narrator intends to transition into a real robbery. To solve the spiraling consequences of this realization, he proposes that following the robbery, they put all the "staff" and enactors on a plane and arrange for it to be blown up, while he and the narrator charter a private plane for their getaway.

Of course, the perfection of which the narrator dreams remains forever out of reach, because there will inevitably be a "residual" (a word the short councillor uses and the narrator has Naz look up in a dictionary): a remainder. Significantly, the residual that causes the enacted-real bank robbery to go terribly wrong is not a presence but an absence. In their practice runs, the enactors had encountered a kink in the carpet that the narrator, with characteristic precision, wanted reproduced exactly, even causing a wood splinter to be inserted under the carpet to make sure the kink would be there. In the enacted-real robbery, however, there was no kink; "the carpet was flat" (290). The narrator describes what happens: "I saw his foot feel for the kink, and feel more, staying behind while the rest of him moved on. The rest of him moved so far on that eventually it yanked the foot up in the air behind it" (290). This "ghost kink" then sets in motion a chain of horrific events, starting with one of the robber enactors accidentally shooting another, then to the narrator reproducing the event in the deserted airplane hangar to which they flee after their getaway from the bank. He characterizes his action as "half instinctive, a reflex," but then admits, "I'd be lying if I said it was only that that made me pull the trigger and shoot [the surviving robber enactor]. I did it because I wanted to" (299).

The "ghost kink" can be understood as the presence of an absence. In larger terms, it functions as more than the initiating wrinkle for the final sequence of events. Throughout the narrative looms another kind of present absence, nonconscious cognition. This is the absence that initiates the narrator's feeling of inauthenticity, the absent functionalities for which consciousness tries to compensate, the missing processes that compel the narrator to more and more extreme measures to feel real. In this sense, the eponymous "remainder" may signify not only the resistance of intractable matter but also the intractable matter of resistance, that is to say, consciousness itself. Bereft of its support system, consciousness refuses to recognize its status as a remainder of a cognitive whole fractured beyond repair, desperately trying to make up for the absence that is never mentioned but whose ghost presence nevertheless dominates this text: the cognitive nonconscious.

BLINDSIGHT AND NEUROSCIENCE

Whereas in *Remainder* the neuroscience references are for the most part implicit, in *Blindsight* they are very much on display. Indeed, Watts on occasion resorts to "infodumps," thinly motivated explanations aimed to enlighten readers; he even includes a bibliography of neuroscientific works he consulted. It is no surprise, then, to find that he includes a character reminiscent of *Remainder*'s narrator, the "synthesist" Siri Keeton. Whereas in *Remainder* the neurological damage results from an accident, with Siri it derives from a radical hemispherectomy, the surgical removal of a hemisphere, an extreme cure for his out-of-control epilepsy. As a result, he is unable to feel empathy and operates largely through rational calculation, as the opening scene reveals when, as a grade school (postoperative) kid, he decides to come to the aid of a friend, Robert (Rob) Paglino, being beaten up by playground bullies. Taking the bullies by surprise, he pulverizes them, unconcerned with the damage he is causing and utterly without empathy for their pain. Even though Rob is the beneficiary of Siri's intervention, he is shocked by its savagery and thereafter calls him "Pod-man" (58).

Like *Remainder's* narrator, Siri responds to his behavioral deficit by crafting compensatory strategies, although of a very different kind. He becomes expert in reading "information topologies," microfacial movements and gestural subtleties communicating intentions, feelings, and motivations that exist independently from semantic content, operating much like the sociometer described in chapter 5 (except that he does it through training his perceptions rather than through an external instrument). Remarking that most people find this disconcerting, he muses, "people simply can't accept that patterns carry their own intelligence, quite apart from the semantic content that clings to their surfaces; if you manipulate the topology correctly, that content just comes along for the ride" (115).

His expertise is the more remarkable because he cannot himself *feel* empathy. He likens his ability to the philosopher John Searle's Chinese Room, which Searle posed as a thought experiment to challenge strong artificial intelligence. Searle imagined a man sitting on a chair in a room with a slot in the door. Someone outside passes a string of Chinese letters through the slot. The man, who does not read or speak any Chinese languages, pulls Chinese characters from a basket at his feet, using a rule book to match the incoming string with another, which he passes through the slot. So convincing are his replies that his interlocutor is convinced the room's occupant understands the strings he composes as answers. But Searle's point is that the man is like a computer; it can match symbols according to given rules, but it has no comprehension of the strings' *meaning*. There have been many responses to the Chinese Room challenge; Siri adopts one of the most compelling, that it is not the man by himself that understands Chinese but the entire room, including the rule book, basket of characters, and even the chair on which he sits.

By analogy, Siri does not understand empathy through mirror neurons or other neurological capacities, but his training and experience (his protocols) enable him to observe with minute care, draw inferences, and extrapolate from these data to conclusions about how someone feels. He comments to his friend Rob, "Empathy's not so much about imagining how the other guy feels. It's more about imagining how *you'd* feel in the same place" (234). By contrast, his method is to read how someone feels and then try to imagine his or her motivations. "I just observe, that's all," he tells Rob. "I watch what people do, and then I imagine what would make them do that" (233).

In this conversation, Rob brings up cases similar to the "Disembodied Woman" from Oliver Sacks (without mentioning Sacks [1998], 43–54), commenting that "Some of them said they felt *pithed.* They'd send a motor signal to the hand and just have to take it on faith it arrived. So they'd use vision to compensate; they wouldn't feel where the hand was so they'd *look* at it while it moved, use sight as a substitute for the normal force-feedback you and I take for granted" (233). He continues, "You use your *Chinese room* the way they used vision. You've invented empathy, almost from scratch, and in some ways—not *all* obviously, or I wouldn't have to tell you this—yours is better than the original. It's why you're so good at Synthesis" (233).

As Rob's comments imply, there remain crucial differences between *feeling* empathy and *re-creating* the knowledge that empathy bestows. Occasionally Siri dreams of his former self, and when he does, he is struck by the vividness of that former life. "I—I dream about him sometimes . . . About . . . *being* him . . . It was—colorful. Everything was more saturated, you know? Sounds, smells. Richer than life" (234). This difference, the feeling of being immersed in a rich sensory environment versus re-creating it from the outside, is the Derridean *différance* between authenticity and reconstruction that wreaked so much havoc in the life of *Remainder's* narrator. For Siri, the deficit is less immediately disastrous, although at crucial moments it surfaces as a decisive force sending him down one path rather than another.

MODIFYING HUMAN (AND NONHUMAN) CONSCIOUSNESS

Watts, who earned a PhD in marine biology, obviously already had a good basic understanding of evolutionary biology, especially with nonhuman species, before he began researching human neuroscience. He uses this background knowledge as well as his research to imagine a cast of characters very far from "normal" psychic functioning. Before setting them in action, he provides the motive for their interactions. An alien species has visited earth vicariously on February 13, 2082, visually capturing earth in a network of optical surveillance devices that simultaneously image every square meter of surface before burning out, as if a global network of flash bulbs had taken earth's picture; people call them the Fireflies. In response, the governments of earth determine to mount an expedition to find the aliens. They discover their location almost accidentally, by intercepting radio signals originating from a comet beyond Pluto in the Kuiper belt. They then put together the team that will man the spaceship *Theseus* in an exploratory journey to make contact with the aliens.

Each crew member represents a specific attempt to push human neurology beyond traditional boundaries. Amanda Bates, augmented with carboplatinum musculature, is the military person trained to command robot warriors, manufactured as needed by the ship's fabricators. Susan James is the linguist who not only has nonsentient brain implants but also had her brain partitioned into four separate areas, each inhabited by a distinct personality. Isaac Szpindel is the biologist who has had his appendages so modified by prostheses that he "heard X-rays and saw in shades of ultrasound" (105), experiencing his lab equipment synesthetically. Siri, the Synthesist, has been sent along as an "objective" observer charged with sending reports back to earth. Finally, the commander is Jukka Sarasti—a vampire.

Watts includes a section on vampire physiology and evolutionary history in the final "Notes and References" section (367–84), noting their superior analytical skills and heightened pattern detection (omnisavantism), superior vision and hearing, and their general cognitive superiority to humans. He also imagines that "vampires lost the ability to code for Υ-Protocadherin Y, whose genes are found exclusively on the hominid Y chromosome" (368), thus making human prey an essential component of their diet. To avoid depleting their food source, they developed the dormancy ("undead") state, which allows them to have extended periods of inactivity. Their evolutionary Achilles' heel is the "Crucifix glitch," a "cross-wiring of normally distinct receptor arrays in the visual cortex" (369) that sends them into grand mal seizures whenever they encounter an array of right angles (which almost never appear in nature but emerge when humans discover Euclidean geometry). For this reason (and only this reason), vampires went extinct until they were genetically reconstituted (a la *Jurassic Park*) by humans late in the twenty-first century.

Including a vampire among the crew allows Watts to render the idea of an ecological niche vividly real, for humans and vampires compete for the same terrestrial niche of high cognitive functioning. Having eradicated, domesticated, or confined to reservations virtually all the other mammal contenders for their niche, humans reign supreme at the top of the food chain. Now that vampires have been resurrected, however, humans who come in contact with them are put in the rare position of being eyed as prey by a superior predator. Even if the constraints of civilization prevent the resurrected vampires from actually eating humans, this is a veneer covering over, but never entirely suppressing, the deep evolutionary history during which humans were helpless to avoid vampire predation. Even with all the modifications, adaptations, and prostheses through which the humans aboard have extended their abilities, the vampire could kill them all in seconds—a fact of which they are uneasily aware as they accept him as their commander. Behind the vampire lurks another presence, the "Quantical AI" computer that runs the ship. "Sarasti was the official intermediary," Siri recounts. "When the ship did speak, it spoke to him—and Sarasti called it *Captain,*" adding, "So did we all" (26).

INTERPRETING THE *RORSCHACH*

With his cast of characters in place, Watts sets the plot in motion by having them encounter the alien ship, self-named the *Rorschach,* cloaked in the shadow of Big Ben, a planet with a gravitational mass of ten Jupiters. Almost thirty kilometers wide, the ship is "not just a torus but a *tangle,* a city-size chaos of spun glass, loops and bridges and attenuate spires" (108). It has an interior environment violently hostile to humans, with a magnetic field "thousands of times stronger than the sun's" (109) and electromagnetic radiation strong enough to be lethal to humans within minutes. Moreover, it is surrounded by aircraft shoveling debris from Ben's accretion belt toward the ship: "particles that collided with the artifact simply *stuck*; *Rorschach* engulfed prey like some vast metastatic amoeba . . . The procession never stopped. *Rorschach* was insatiable" (109). The implication, of course, is that *Rorschach* is still growing; when it reaches maturity, it will become even more formidable, designed for purposes that the humans fear will include annihilation of earth.

Sarasti reasons that there is no time to waste and orders his crew to carry out forays inside the *Rorschach* and, if possible, capture specimens. Amanda Bates, tactical commander of the *Rorschach* incursions, orders the others into a shielding tent while remaining outside herself. When her fellow crew members try to communicate with her, however, she utters enigmatic responses, saying, "I'm dead already" (162), "I'm not out here," "[I'm] nowhere," "I'm nothing" (171). Siri, looking at her faceplate, remarks, "I could tell that something was missing. All her surfaces had just *disappeared*" (162). Later, Szpindel tells Siri that Bates did not just "believe" she did not exist but "*knew*" it. For a fact" (180). Siri, intrigued, consults ConSensus, the ship's equivalent of an interactive encyclopedia, to look up all the ways in which traumatized brains cause radical misperceptions of the body and outside world (here one of those infodumps appears, 193). Much later, he summarizes his conclusions: "The brain stem does its best. It sees the danger, hijacks the body, reacts a hundred times faster than that fat old man sitting in the CEO's office upstairs; but every generation it gets harder to work around this—this creaking neurological bureaucracy" (302).

The reference to knowledge that the brainstem has but cannot communicate to the "creaking neurological bureaucracy" "upstairs" is Watts's version of the neurological phenomenon known as blindsight, inspiration for the book's title. Lawrence Weiskrantz of Oxford University, a prominent researcher into blindsight, tracked for ten years a patient who had a small tumor removed from the visual center V1 on one side of the brain, leaving him blind to anything that happened on his other side (because of the crossover of neural connections) (see, for example, Weiskrantz et al. 1974). The patient reported that when events happened rapidly, he somehow knew something was occurring even though he could not see it. Max Velmans explains the phenomenon:

Blindsight [is] a condition in which subjects are rendered blind in one half of their visual field as a result of unilateral striate cortex damage. If stimuli are projected to their blind hemifield subjects cannot see them in spite of the fact that their full attention is devoted to the task. As they cannot see the stimulus they maintain that they have no knowledge about it. However, if they are persuaded to make a guess about the nature of the stimulus in a forced choice task, their performance may be very accurate. For example, one subject investigated by Weiskrantz et al. (1974) was able to discriminate horizontal from vertical stripes on 30 out of 30 occasions although he could not see them. In short, the subject *has* the necessary knowledge but *does not know that he knows.* In information processing terms, it is as if one (modular) part of his system has information which is not generally available throughout the system (Velmans 1995).

The alien life-form encountered by the *Theseus* crew, the so-called scramblers, are like people with blindsight in that they know but do not know that they know, because they lack consciousness. Watts has his characters speculate that compared to the scramblers, consciousness as it evolved in humans carries with it costs that impose heavy evolutionary penalties. "It wastes energy and processing power, self-obsesses to the point of psychoses. Scramblers have no need of it, scramblers are more parsimonious. With simpler biochemistries, with smaller brains—deprived of tools, of their ship, even of parts of their own metabolism—they think rings around you . . . they turn your own cognition against itself. They travel between the stars. This is what intelligence can do, unhampered by self-awareness" (302).

Bates, although she went into *Rorschach* as a human with a self, experiences a radical loss of self in its violently alien environment, temporarily becoming much closer to the scramblers. She literally becomes "no one,"[3] reduced to nonconscious cognitive and material processes that precede the self's construction by consciousness. Her pronouncement "I'm dead" signals not the end of organismic life but the cessation of the narrating self, the "I" whose topological surfaces are wiped clean when consciousness ceases to function. When Siri understands, this is how he parses her situation: "for Amanda Bates to say 'I do not exist' would be nonsense, but when the [nonconscious] processes beneath say the same thing, they are merely reporting that the parasites [conscious processes] have died. They are only saying that they are free" (304).

Although Bates recovers her sense of self when she returns to *Theseus,* her consciousness's erasure more than hints that the aliens native to *Rorschach's* environment function without consciousness as well. When the humans bring back a dead specimen, they submit it to extensive testing. The alien scramblers have neuronal structures utterly unlike humans. Robert Cunningham (the biologist wakened from space hibernation when Szpindel is killed in an incursion) announces that the scramblers have "no cephalization, not even clustered sense organs. The body's covered with something like eyespots, or chromatophores, or both . . . Every one of those structures is under independent control . . . The entire body acts as a single diffuse retina. In theory that gives it enormous visual acuity" (224). He pronounces the alien "an absolute miracle of evolutionary engineering," but because there is no central nervous system, he also judges it to be "dumb as a stick" (226).

When the humans capture live specimens on their final incursion, testing reveals how wrong this judgment is.[4] At first unresponsive, the scramblers are detected communicating covertly, and the humans decide to hurt/torture them to force cooperation. Thereupon they show remarkable geometric skills, and when prodded with number sequences, "were predicting ten-digit prime numbers on demand" (265). When Cunningham dismisses these as "splinter skills," Susan James draws the obvious conclusion: "They're *intelligent,* Robert. They're smarter than us. Maybe they're smarter than *Jukka* [the vampire]. And we're—Why can't you just *admit* it?" (265). Later she says to Siri, "They're intelligent; we know they are. But it's almost as though they don't know they know, unless you hurt them. As if they've got blindsight spread over every sense" (274).[5]

BLINDSIGHT AND THE COSTS OF CONSCIOUSNESS

The eponymous blindsight appears sporadically in the text, for example when Szpindel explains to Siri how he almost caught the battery tossed to him although his conscious visual perception was blocked by *Rorschach's* electromagnetic field. "Nothing wrong with the receptors . . . Brain processes the image but it can't access it. Brain stem takes over" (170). Later he elaborates: "You just—you get a feeling, is all. A sense of where to reach. One part of the brain playing charades with another, eh?" (180). Blindsight functions as a synecdoche for all the trauma-induced and brain-damaged syndromes mentioned in infodumps and explained in the "Notes and References" section: in different ways, all reveal the limitations of conscious thought and the inadequacy of equating cognition with consciousness alone.

Watts postulates that for aliens who have never developed consciousness, the parts of human languages that describe conscious perceptions, feelings, and responses would be so much noise. "Imagine you are a scrambler," Siri says. "Imagine you have intellect but no insight, agenda but no awareness. Your circuitry hums with strategies for survival and persistence, flexible, intelligent, even technological—but no other circuitry monitors it. You can think of anything, yet are conscious of nothing" (323). Then, he continues, imagine how human language would sound to you. "The only explanation is that something has coded nonsense in the way that poses as a useful message; only after wasting time and effort does the deception become apparent. The signal functions to consume the resources of a recipient for zero payoff and reduced fitness. The signal is a virus. Viruses do not arise from kin, symbionts, or other allies. The signal is an attack" (324). He concludes there can be no rapprochement between humans and this alien species; "How do you say *We come in peace* when the very words are an act of war?" (325). The very fact that human language springs from consciousness ensures that the aliens see humans as evolutionary enemies.

As the costs of consciousness become more widely discussed in contemporary culture (for example, Hansen 2015), other writers also imagine language not as the singular achievement of the human species, as commentators such as Steven Pinker would have it, but as a virus, a disease, an evolutionary hiccup about to be replaced by other modes of communication, a trajectory pioneered by William Burroughs in *Naked Lunch* (1959) and other works. In *The Flame Alphabet* (2012), for example, Ben Marcus writes of children's language toxic to their parents, sickening them the more they listen to it, and eventually becoming lethal. In *The Silent History,* Eli Horowitz, Matthew Derby, and Kevin Moffett (2014) imagine a generation of children who do not understand verbal language and are unable to learn it, no matter how desperately their parents strive to inculcate it. Instead they communicate with one another through microfacial gestures, developing an entire vocabulary that the children learn quickly and easily and use to communicate among themselves. One of the narrators, witnessing this communication, asks, "What unknown abilities had filled this void [the absence of verbal language]? Was the world somehow brighter, more tangible, without the nagging interference of language? Was the absence of words actually a form of freedom?" (8). We may extend these questions to consciousness itself. Without words, would the narrating consciousness be silenced? If so, would the self be configured in new ways or disappear altogether? The very idea that language could be a liability rather than a crowning achievement illustrates how the costs of consciousness may be invoked to question the basis for human exceptionalism and the privileges it has traditionally bestowed, including the Baconian imperative for humans to achieve dominion over all other creatures on the earth, and over the earth itself.

ADVANCED TECHNOLOGY WITHOUT CONSCIOUSNESS

One of the challenges Watts faces is how to account for the aliens' vastly superior technology, achieved without conscious awareness. He locates the cause in emergent complexity. Michael Dyer, a computer scientist specializing in artificial intelligence and my former colleague at UCLA, remarked that the more intelligent the environment, the less intelligence one needs to put in the heads of the agents in an artificial life simulation, because the environment's structured specificities make it possible for the agents to evolve emergent complexities through their interactions with it. Cunningham, *Theseus's* biologist, proposes a similar dynamic for the alien ship *Rorschach* and the scramblers. To explain, he instances the well-known example of a honeycomb. No bee has an overall plan for the honeycomb in its head; all it has is an instinct to turn in a circle and spit wax while adjacent bees do the same. The wax lines press against each other to form a hexagon, the polygon with the closest packing ratio, and the honeycomb is the emergent result. In the case of *Rorschach,* Cunningham observes, it is the *scramblers* who are the honeycomb, the emergent result of the dynamics of the ship's environment (267). "I don't think *Rorschach's* magnetic fields are counterintrusion mechanisms at all. I think they're part of the life support system. I think they mediate and regulate a good chunk of scrambler metabolism" (267). The aliens represent, then, not only distributed cognition but distributed organismic life.

Unlike on earth, where freely living independent organisms developed first and then created technology, here the technology and biological life evolved together, each stimulating the other to deeper interactions and greater emergent complexity.[6] Cunningham discovers that the scramblers have no genes and no independent reproductive mechanism; the ship grows them in a stack, each scrambler with a navel in front and behind, and the top one buds off and becomes mobile when it is mature. Devoid of consciousness yet still intelligent, the scramblers have no selves to create an imperative to survive; rather, the drive to survive is invested in ship-plus-scramblers, for without the ship environment, the scramblers would die within days because their metabolism slowly decays without the ship's environment to regulate and replenish it. When they are captured and taken into *Theseus,* Cunningham remarks they are metaphorically "holding their breath. And they can't hold it forever" (267).

In contradistinction to the scramblers, the mainstream view of humans imagines us first as independent (and social) organisms, and our technologies as late-addition cultural achievements—nice to have, certainly, but hardly intrinsic to our survival. However, if all technical cognitive systems were to bite the dust tomorrow, the result would be systemic chaos and a massive die-off of the human species. Imagine all transportation systems inoperative (even cars and trucks have computerized ignition systems, and railroads and airplanes are completely interpenetrated by computational devices), all water and sanitation facilities off-line, all electric grids down, all national and international supply lines cut, banking systems crashed, agricultural and livestock production at a standstill, all medical equipment except the most robust hand instruments unavailable, etc. No doubt some people in rural and remote areas would survive, but the death tolls would likely mount to millions or billions. Why do we continue to think of ourselves as beings independent from our technologies and capable of living without them? Here we may recall Damasio's insight: "consciousness . . . constrains the world of the imagination to be first and foremost about the individual, about an individual organism, about the self in the broad sense of the term" (2000, 300). Consciousness, that is, insists that the human self is the primary actor, and technologies mere prostheses added on later (compare with Stiegler [1998]).

Watts gestures toward the deep interconnection between technical and human cognition in his climax, when Sarasti prepares to make *Theseus* itself into a weapon and dive it into the *Rorschach,* condemning the crew to instant death—except for Siri, who is instructed to escape in a probe and return to earth to warn people about the alien threat. Susan James rebels, however, and tampers with Sarasti's anti-Euclidean drugs. As a result he goes into a grand mal seizure—whereupon one of the robot warriors crushes his skull and inserts some electronics, and the undead body begins moving again, now under the control of the Captain, the "Quantical AI" at the ship's heart. Siri, unnerved, demands to know if Sarasti ever was the commander:

"Did he ever speak for himself? Did he decide *anything* on his own? Were we ever following *his* orders, or was it just you all along?" (353). Tapping on a keyboard the undead body has seized, the Captain gives him his answer: "U DISLKE ORDRS FRM MCHNES. HAPPIER THIS WAY" (353). It is ironic, of course, that the humans are imagined as happier taking orders from a vampire, their evolutionary archenemy, than from a computer. From the perspective of their dependence on the ship's AI, the difference between them and the scramblers is not quite so great as it seemed—or more accurately, not as enormous as consciousness imagined it to be.

As Siri begins the long fourteen-year journey home, he meditates on the role of humans in the universe. Susan James, resisting Sarasti's implication that consciousness is a disability, asks why, in that case, had humans survived: "If [consciousness] were really so pernicious, natural selection would have weeded it out" (306). Sarasti delivers the riposte: "You have such a naïve understanding of evolutionary processes. There's no such thing as *survival of the fittest*. *Survival of the most adequate,* maybe. It doesn't matter whether a solution's optimal. All that matters is whether it beats the alternatives" (306).

Interpreting this observation in topological terms, we may imagine a fitness landscape in which a local maximum has arisen. Elsewhere, however, an even larger global maximum towers. The problem from an evolutionary viewpoint is that an organism perched on the local maximum can never reach the global one, because to get there, it would have to go downhill, becoming less fit, to traverse the intervening distance. Consequently, it is more fit relative to its immediate competitors, but not from a global perspective. Earth in this analogy is the local environment, and on it, humans have achieved a local maximum within their ecological niche (leaving aside the mythical vampires). Elsewhere in the universe, however, a larger global maximum rules. Siri had earlier grasped Sarasti's implication: "scramblers were the norm. Evolution across the universe was nothing but the endless proliferation of automatic, organized complexity, a vast arid Turing machine full of self-replicating machinery forever unaware of its own existence. And we—we were the flukes and the fossils" (325).

As he monitors communications streaming from earth, he begins to hear less music, and less human speech. He also receives a "general delivery" communication from his father that he interprets as a coded warning not to return. As the language coming from earth turns from words to the clicks and hisses characteristic of vampires, he begins to suspect that, like the dinosaurs in *Jurassic Park,* they have snapped their chains and broken out of their enclosures, now running rampant through the world. Moreover, he suspects that they have begun to evolve away from consciousness into non-conscious modes of being. "We humans were never meant to inherit the Earth," he muses. "Vampires were. They must have been sentient to some degree, but that semi-aware dreamstate [in which they live] would have been a rudimentary thing next to our own self-obsession. They were weeding it out. It was just a phase" (362). Contemplating how he has changed, he says, "Thanks to a vampire and a boatload of freaks and an invading alien horde, I'm Human again. Maybe the last Human. By the time I get home, I could be the only sentient being in the universe" (362). Consciousness, in this view, was a clumsy evolutionary work-around that was better than its immediate competitors, but in the long run, cost more than it was worth, and was weeded out when the context for competition widened beyond earth to the universe.

What are we readers, conscious beings all, to make of this conclusion?

"NORMAL" CONSCIOUSNESS AND TECHNICAL COGNITION

One way to evaluate these texts is to analyze the strategies they use to construct normality (the place where presumably we readers are). Viewed from this angle, strategies radically different from each other come into sight. In *Remainder,* the narrator starts from a place with which we can easily identify. He has suffered a serious accident through no fault of his own, an innocent bystander; as a result, he has had to struggle to regain functionality. From here, of course, he begins his disastrous slide into obsession and then psychosis, as his need to control spirals outward into the world. In *Blindsight,* by contrast, the protagonists begin very far from the human baseline, sporting technical modifications and augmentations in nearly every way imaginable— and then there's the vampire, creature of myth and Gothic fiction. As the narrative progresses, however, they come to seem normal by comparison with the aliens, their differences from unmodified humans swamped by the overwhelming tsunami of difference that the aliens represent. Then at the end, this normality is flipped again into the abnormal, the fate of humans in a nonconscious universe.

Technical cognition also plays very different roles in the two texts. In *Remainder,* it is conspicuously absent. Absurdly by today's standards, the narrator, when he encounters a word or concept he doesn't know, has Naz *call a colleague in the office,* who looks it up, calls Naz back, and Naz then relates it to the narrator. Published in 2005, *Remainder* was already well into the era of the Palm Pilot and Blackberry, introduced in 2001. Granted that the now-ubiquitous iPhone was not marketed until 2007, there were plenty of smartphones and digital handheld devices already around, so this procedure cannot simply be business as usual. Rather, it seems designed to call attention to the fact that technical cognition plays virtually no role in the text. When the narrator wants to enlist reenactors, he chooses humans to do his bidding. He operates not within a distributed cognitive network but a self-designed, singularly idiosyncratic network to extend his own (dysfunctional) cognitions into the world.

Moreover, as we have seen, he uses this network to impose his will on reality. A small indication of this occurs when the (hallucinated) short councillor uses the word "residual" (259). The narrator has Naz do the cell phone routine to look up the word, spelling it out for him. "A noun," Naz pronounces, and then asks, "What short councillor?" Repeating the councillor's words as he heard them, the narrator continues, "This strange, pointless residual. And he pronounced the *s* as an *s,* not a *z.* Re-*c*-idual. Have it looked up with that spelling." Since this pronunciation is an artifact of his hallucination, it naturally does not exist in common usage: "Word not found," Naz reports. At this, the narrator becomes enraged, commanding Naz to "tell them to go and find a bigger dictionary, then!" "I was feeling really bad now," he says, continuing, "And if you see that short councillor here . . . ," whereupon Naz again asks, "What short councillor?" (271). Played out at length, the scene illustrates the narrator's belief that *he* can determine what counts as reality, including what words appear in a dictionary. Were technical cognition to play a larger role in the text, the narrator would be confronted with this kind of situation all the time, as technical systems are immune to bribes and payoffs. By restricting his network to humans, the narrator is able to persist in his delusions.

In *Blindsight,* by contrast, the interpenetration of human cognition with technical cognitive systems constitutes the "new normal" in this fictional future (2082 and beyond). Like *Remainder's* narrator, Siri has suffered a devastating intervention and struggles to regain normalcy, an aspiration with which readers can identify. That he never entirely succeeds is illustrated in the scene where his sometime lover, Chelsea, contacts him because she has contracted a deadly virus and wants to see him again, reaching out before she dies. The "normal" reaction would of course be to call her back immediately. Instead, Siri spends days searching for the right "algorithm that fit" (294), the rules that will tell him how to act, what to say. The effect is not to see him as a monster, however, as is the case with *Remainder's* narrator, but rather sadness that in this difficult situation, he is unable to act at all.

As we come to know the protagonists aboard *Theseus,* the estrangement effect diminishes, and their reactions, notwithstanding their technical enhancements and modified cognitions, seem very familiar. When the scramblers are introduced, the crew's differences from the presumed normality of readers recedes into insignificance. But when Siri on the voyage home thinks he may be the last sentient creature in the universe, that normality is suddenly converted into abnormality. He has warned us readers repeatedly that he is not representing what literally happened aboard the *Theseus*, or what the characters actually said, but rather what they meant. His final words drive the message home: "So I really can't tell you one way or the other. You'll just have to imagine you're Siri Keeton" (362), inviting us to identify with him as one dodo identifies with another. Technical cognition, no matter how advanced, cannot remedy this situation, because consciousness is always part of the mix, calling the shots, designing the hardware, determining how and in what ways the interfaces work. Unless, of course, we consider Sarasti and the Captain, in which case consciousness again becomes the outlier in a universe dominated by the nonconscious.

What these texts demonstrate, then, is how the very idea of "normal" mutates when the costs of consciousness are taken into account. They suggest that "normality" cannot be sufficiently anchored by consciousness alone, or indeed by human cognition. In a universe where technical cognition is already on the scene and NASA announces the discovery of earth-like planets in other solar systems,[7] human cognition can no longer be regarded as the "normal" standard against which all other cognition can be measured, technical and nonhuman, terrestrial and alien. If decentering the human is a major thrust of contemporary cultural theory, including animal studies, posthumanities, new materialisms, and other such projects, the entire basis for cognition shifts to a planetary scale, in which human actors are but one component of complex interactions that include many other cognizers. Whether consciousness is a crown or a burden, or both together, must be reevaluated in this larger context of planetary cognitive ecology— and perhaps beyond planetary as well.

PART 2

COGNITIVE ASSEMBLAGES

CHAPTER 5

Cognitive Assemblages: Technical Agency and Human Interactions

In a passage from *Reassembling the Social: An Introduction to Actor-Network-Theory* (2007), Bruno Latour criticizes what he calls the "sociology of the social" for making an artificial distinction between humans and objects with this example. "Any human course of action might weave together in a manner of minutes, for instance, a shouted order to lay a brick, the chemical connection of cement with water, the force of a pulley unto a rope with a movement of the hand, the strike of a match to light a cigarette offered by a co-worker" (74). While I very much admire Latour's work and happily acknowledge the significant contributions of actor-network-theory (ANT) to science studies, this passage illustrates why a framework focusing on cognition adds an important dimension to existing approaches to complex human systems. Notice that the action begins with a "shouted order," and that material forces are then enlisted as a result of this decision. The cement could not by itself build a structure; for that matter, cement relies on human intervention to come into existence as a construction material. In short, cognitive processes play crucial roles in Latour's example, notwithstanding that he intends it to show the symmetry between human actions and material forces. The point, as I have emphasized, is not to glorify human choice but rather to expand the spectrum of decision makers to include all biological life-forms and many technical systems. Decision makers certainly can and do enlist material forces as their allies, but they are the ones who try to steer the ship in a particular direction.

The term I use to describe these complex interactions between human and nonhuman cognizers and their abilities to enlist material forces is "cognitive assemblage." While Latour and Deleuze and

Guattari (1987) also invoke "assemblage," a *cognitive* assemblage has distinctive properties not present in how they use the term. In particular, a cognitive assemblage emphasizes the flow of information through a system and the choices and decisions that create, modify, and interpret the flow. While a cognitive assemblage may include material agents and forces (and almost always does so), it is the cognizers within the assemblage that enlist these affordances and direct their powers to act in complex situations.

Having invoked the idea of power (most tellingly in the book's title, and in passing throughout), I will now indicate how power— and its handmaiden, politics— appear in this framework. Here Latour offers valuable guidance, for he points out that power is an effect produced by mediators (human and nonhuman) that transform temporary and shifting configurations into durable, robust, and reproducible structures capable of creating, solidifying, and wielding power. Responding to critiques of ANT as a theory that ignores politics and social inequalities, Latour argues that its focus on mediators is precisely what enables politics to come into view as provisional practices that can always be otherwise than they are. "Sociologists of the social," as Latour characterizes his colleagues who invoke "the social" as an explanatory force, ignore the mediators that make power possible, thereby mystifying power so that constructive change becomes more difficult to imagine or initiate. "By putting aside the practical means, that is the mediators, through which inertia, durability, asymmetry, extension, domination are produced and by conflating all those different means with the powerless power of social inertia, sociologists, when they are not careful in their use of social explanations, are the ones who hide the real causes of social inequalities" (85).

Although Latour would not agree with the distinction I make between cognizers and material processes, his focus on mediators fits well with my vision of cognizers as transformative actors. By reenvisioning cognition and crafting a framework in which nonconscious cognition plays a prominent role, my approach enables analyses of cognitive assemblages, and the mediators operating within them, as the means by which power is created, extended, modified, and exercised in technologically developed societies.

The charge that Latour levels at sociology can apply equally (or even more so) to the humanities. "To the *studied* and *modifiable* skein of means to achieve powers, sociology, and especially critical sociology, has too often substituted an invisible, immoveable, and homogeneous world of power for itself . . . Thus, the accusation of forgetting 'power relations' and 'social inequalities' should be placed squarely at the door of the sociologists of the social" (86). His emphasis here on the *studied* and *modifiable* implies that modification of power relations requires detailed and precise analyses of the ways in which assemblages (in my terms, cognitive assemblages) come together, create connections between human and technical actors, initiate, modify, and transform information flows, thereby bringing contexts into existence that always already determine the kinds and scope of decisions possible within milieus and the meanings that emerge within them.

On the offensive, Latour drives home his critique of critical sociology in strong terms that, unfortunately, are resonant with the critical humanities as well. "If, as the saying goes, absolute power corrupts absolutely, then gratuitous use of the concept of power by so many critical theorists has corrupted them absolutely—or at least rendered their discipline redundant and their politics impotent" (85). This last phrase stings with special force because it is so obviously true of the critique of power as it was displayed in the humanities during the 1970s, 1980s, and beyond. If we judge a political agenda by its efficacy in persuading a populace, then the deconstructive theory that swept the humanities during this period had, in a generous interpretation, mixed results: while it succeeded in radicalizing many academics within the humanities, it estranged the humanities from the general public with discourses seen as obscure, not to mention nonsensical, and made the humanities seem increasingly peripheral to the main business of society. This suggests that at the very least, it might be time to try another approach that analyzes power relations by focusing on how power is created, transformed, distributed, and exercised in an era when complex human systems are interpenetrated by technical cognition—that is to say, by focusing on cognitive assemblages.

I turn now to parsing this key term in more detail. In Deleuze and Guattari's usage, "assemblage" [*agencement*] carries the connotations of connection, event, transformation, and becoming. They privilege desire, affect, and transversal energies over cognition, but the broader definition of "cognition" that I employ brings my argument somewhat closer to theirs, although significant differences remain. I want to convey the sense of a provisional collection of parts in constant flux as some are added and others lost. The parts are not so tightly bound that transformations are inhibited and not so loosely connected that information cannot flow between parts. An important connotation is the implication that arrangements can scale up, progressing from very low-level choices into higher levels of cognition and consequently decisions affecting larger areas of concern.

In focusing on cognition, which the reader will recall I defined as "a process of interpreting information in contexts that connect it with meaning," I highlighted the activities of interpretation, choice, and decision and discussed the special properties that cognition bestows, namely flexibility, adaptability, and evolvability. A cognitive assemblage approach considers these properties from a systemic perspective as an arrangement of systems, subsystems, and individual actors through which information flows, effecting transformations through the interpretive activities of cognizers operating upon the flows. A cognitive assemblage operates at multiple levels and sites, transforming and mutating as conditions and contexts change.

Why choose assemblages rather than networks, the obvious alternative? The question is especially pertinent, since "network" is usually favored by Latour (witness ANT), although he tends at times to use "assemblage" as a synonym (Latour 2007). Networks are typically considered to consist of edges and nodes analyzed through graph theory, conveying a sense of sparse, clean materiality (Galloway and Thacker 2007). Assemblages, by contrast, allow for contiguity in a fleshly sense, touching, incorporating, repelling, mutating. When analyzed as dynamic systems, networks are like assemblages in that they function as sites of exchange, transformation, and dissemination, but they lack the sense of those interactions occurring across complex three-dimensional topologies, whereas assemblages include information transactions across convoluted and involuted surfaces, with multiple volumetric entities interacting with many conspecifics simultaneously.

Because humans and technical systems in a cognitive assemblage are interconnected, the cognitive decisions of each affect the others, with interactions occurring across the full range of human cognition, including consciousness/unconscious, the cognitive nonconscious, and the sensory/perceptual systems that send signals to the central nervous system. Moreover, human decisions and interpretations interact with the technical systems, sometimes decisively affecting the contexts in which they operate. As a whole, a cognitive assemblage performs the functions identified with cognition in general: flexibly responding to new situations, incorporating this knowledge into adaptive strategies, and evolving through experience to create new strategies and kinds of responses. Because the boundaries are fuzzy, where one draws the line often depends on the analytical perspective one uses and the purposes of the analysis. Nevertheless, for a given situation, it is possible to specify the kinds of cognitions involved and consequently to trace their effects through an evolutionary trajectory.

The most transformative technologies of the later twentieth century have been cognitive assemblages: the Internet is a prime example. While many modern technologies also had immense effects—the steam engine, railroads, antibiotics, nuclear weapons and energy—cognitive assemblages are distinct because their transformative potentials are enabled, extended, and supported by flows of information, and consequently cognitions between human and technical participants. Hybrid by nature, they raise questions about how agency is distributed among cognizers, how and in what ways actors contribute to systemic dynamics, and consequently how responsibilities—technical, social, legal, ethical—should be apportioned. They invite ethical inquiries that recognize the importance of technical mediations, adopting systemic and relational perspectives rather than an emphasis (I would say overemphasis) on individual responsibility.

Developing the concept of a cognitive assemblage in this chapter, I begin with the basic level of a city's infrastructure. Nigel Thrift (2004) has called infrastructures governing our everyday assumptions about how the world works the "technological unconscious," consisting of predispositions that regulate our actions in unconscious and nonconscious ways through routine anticipations, habitual responses, pattern recognition, and other activities characteristic of the cognitive nonconscious. From there my analysis moves inward toward the body to discuss digital assistants that interact directly on a personal level. As these devices become smarter, more wired, and more capable of accessing informational portals throughout the web, they bring about neurological changes in the mindbodies of users, forming flexible assemblages that mutate as information is gathered, processed, communicated, stored, and used for additional learning that affects later interactions. As the user's responses and interactions reveal more and more about her predispositions of which she may not even be aware, the possibility for surveillance grows progressively stronger, a trajectory analyzed here through the sociometer developed by Alex

("Sandy") Pentland (2008) and epitomized by Frans van der Helm's extreme proof-of-concept in the MeMachine (AR Lab 2013).

I turn next to analyze the implications of increasing technical autonomy evident in many research programs now underway, for example self-driving cars, face-recognition systems, and autonomous weapons. My focus is on the transition from pilot-operated drones to autonomous drone swarms. Amplifying technical autonomy requires that the cognitive capabilities of technical devices be increased, so distributed agency is preceded by and dependent on a prior redistribution of cognition. The tendency of technical devices to unsettle discursive formations and shake up cultural practices is exacerbated in the case of military drones, where nothing less than life and death may be at stake. Illustrative sites for such seismic disturbances are international treaties delineating the so-called laws of war, which assume that agency, and consequently decisional power, lie entirely with humans, without considering the effects of technical mediations. The significant changes brought about when technical devices do have agency illustrate what happens to complex human social systems when they are interpenetrated by technical cognition. As I will show, the resulting cognitive assemblages transform the contexts and conditions under which human cognition operates, ultimately affecting what it means to be human in developed societies.

INFRASTRUCTURE AND TECHNICAL COGNITION

Imagining the future of technical cognition, Alex ("Sandy") Pentland of MIT Media Lab writes, "It seems that the human race suddenly has the beginnings of a working nervous system. Like some world-spanning living organism, public health systems, automobile traffic, and emergency and security networks are all becoming intelligent, reactive systems with . . . sensors serving as their eyes and ears" (2008, 98). The analogy has not been lost on neuroscientists, who adopt traffic metaphors to characterize information flowing through the body's nervous system (Neefjes and van der Kant 2014). As Laura Otis argues about the connections nineteenth-century scientists made between nerves and networks of telegraph lines, such analogies have real conceptual force: "metaphors do not 'express' scientists' ideas; they *are* the ideas" (2001, 48).

A good site to explore interactions between a city's "nervous system" and that of humans is ATSAC, the Automated Traffic Surveillance and Control system in Los Angeles that controls traffic on 7,000 miles of surface streets (ATSAC n.d.). I made a site visit there in November 2014 and spoke with Edward Yu, ATSAC's director. The computer system at ATSAC's heart, fed by information flowing from sensors and actuators throughout the city, is flexible, adaptive, and evolutionary, capable of modifying its own operations. Combined with human operators who work with it, ATSAC illustrates the ways in which technical nonconscious cognition works with human capabilities to affect the lives of millions of urban inhabitants.

Located four levels below the street, the center is housed in an underground bunker originally designed to protect city officials from bomb attack (that it has now been turned over to ATSAC is perhaps an unintentional acknowledgment of how crucial traffic control is to Los Angeles). Information flowing into the center includes reports from 18,000 loop detectors, working by magnetic induction, that register traffic volume and speed every second in over 4,000 intersections, while more than 400 surveillance cameras monitor the most troublesome or important intersections. Analyzing these data streams, computer algorithms automatically adjust the signal lights to compensate for congested lanes. This requires changing the relevant signals in coordinated fashion, so that side streets coming into a main thoroughfare, for example, are timed to work together with the main street's lights. The system also monitors traffic in the dedicated bus lanes; if a bus is behind schedule, the algorithms adjust the signals to allow it to make up time. All the monitoring information is processed in real time. The entire system is hardwired to prevent latency, with copper wire from the loop detectors running into hubs, where the information is carried by fiber optic into the center. ATSAC thus represents a considerable civic investment in infrastructure. Begun for the 1984 Olympics in Los Angeles, it was finally completed in 2012.

In addition to everyday traffic, engineers adapt the system for special events such as a presidential visit or a blockbuster premiere. Even the automatic cycles have special provisions. In neighborhoods with large Orthodox Jewish communities, for example, the push buttons operating "walk" signals are programmed to work automatically during the daylight hours of the Jewish Sabbath, during which times Orthodox Jews are prohibited from working machinery and thus cannot operate the buttons manually. Since daylight hours differ by seasons, the system has programmed into it the times for sunrise and sunset for the entire year.

Without the help of the system's sensors, actuators, and algorithms, it would be impossible for humans to conduct such widespread traffic coordination, and prohibitively expensive even to attempt it. According to studies, the system has resulted in 20 to 30 percent fewer stops at intersections, reduced travel time by 13 percent, cut fuel consumption by 12.5 percent, and air emissions by 10 percent (Rowe 2002). These statistics have real consequences for the lives of Angelenos. Having lived in Los Angeles for two decades, I can testify how important traffic patterns become, often dictating life choices such as work schedules, entertainment possibilities, and friendship networks. When Yu attends community meetings, he likes to ask audiences how many have been affected by a major crime. Typically, two or three people out of a hundred raise their hands. Then he asks how many have had their lives affected by traffic; hands shoot up from virtually everyone.

Specifically, how do the technical cognitions instantiated in ATSAC interact with human cognitions? The algorithms are coordinated with a database in which the traffic information is stored for a week; this allows the system to extract patterns, and it uses these patterns to update the algorithms accordingly. Drivers also detect patterns, no doubt at first consciously and then nonconsciously as they travel the same routes over and over. When anomalies occur, they are quick to notice and often call the center to alert operators to problems at particular intersections. The operators also must internalize the patterns so they can make informed decisions. Yu reported that it typically takes about a year and a half to train new personnel before they have enough experience to distinguish the ordinary from the extraordinary. For example, Santa Monica Boulevard feeds into the Santa Monica freeway; if the freeway entrance is backed up, there is no point in arranging the signals to permit traffic to move faster on the street, since that would only exacerbate the problem. When an intersection becomes congested, the screen graph displaying that information begins to flash red, and the operator can pull up the camera feed to identify the problem. The afternoon I visited, an intersection on Alameda Street in the downtown area became congested, and the camera image revealed that police had blocked off a portion of the street to prepare for a demonstration. Unfortunately they had not informed the center, and with rush hour approaching, active intervention was necessary to prevent traffic from snarling throughout the downtown area. With a single command, an operator can change a whole network of signals, as was necessary in this instance.

ATSAC exemplifies productive collaboration between human conscious decisions, human nonconscious recognition of patterns by both operators and drivers, and the technical cognitive nonconscious of the computer algorithms, processors, and database. As Ulrik Ekman notes in discussing the topology of intelligent cities, "Design here must meet an ongoing and exceedingly complex interactivity among environmental, technical, social and personal multiplicities of urban nodes on the move" (2015, 177). Functioning within these complexities, ATSAC demonstrates how a cognitive assemblage works. At any point, components are in flux as some drivers leave the system and others enter. Although the underlying infrastructure is stable, the computer's screenic interfaces constantly change as older trouble spots are cleared up and new ones emerge. Similarly, the basic cognitive structures of the algorithms are set, but they are also modified through the extraction of patterns, which are used to modify continuing operations as contexts change and new meanings are produced.

The political assumption instantiated in the system's cognitive functioning is that it is desirable for traffic to flow smoothly. In this sense, it contributes positively to the experiences of Angelenos. The downside, of course, is that by decreasing traffic congestion, it allows the (in)famous L.A. dependence on cars to continue and even increase. Yet the system has also been engineered to encourage increased use of public transport through the dedicated bus lanes it manages.[1] Consequently, it can be argued that the net results are positive overall. The system is the beneficiary of investments by the city over several decades and different political regimes, which nevertheless managed to summon the political will to keep the system running and enlarge it until the entire city was included. ATSAC thus shows the possibilities for constructive outcomes from the deep penetration of the technical cognitive nonconscious into complex human systems. It is worth noting that the system has no direct connection to market considerations of profit and loss.

DIGITAL ASSISTANTS AND INFORMATION PORTALS

From the mediations of an urban traffic infrastructure, we move up close and personal to consider the more intimate and arguably more neurologically consequential interactions with a digital assistant. VIV, a device being developed by VIV Labs in San Jose, CA, evolves its capacities through web reading, geolocation, mobile interactions, and real-life queries (Levy 2014). The program, soon to be marketed as the "next generation Siri," combines GPS orientation with an open system that programs on the fly, parses sentences, and links to third-party sources. Developers Dan Kittlaus, Adam Chever, and Chris Brigham say that VIV can parse relatively complex queries such as this, shown in *Wired* alongside a flowchart indicating VIV's search techniques: "On the way to my brother's house, I need to pick up some cheap wine that goes well with lasagna." The program first parses "brother" as a family relationship and looks for the appropriate entry in the user's Google contacts. It then plots the route and, noticing a missing variable, asks the user how far she is willing to deviate to get the wine. With this information, the program searches the web for lasagna recipes, identifying it as a food item containing tomato, cheese, and pasta. Searching for wine-food matches yields appropriate varietals, and further inquiries yield price ranges from which a "cheap" spectrum is identified. The next step queries catalogues of the wine stores en route, with the final result being a list of wines at the appropriate stores.

VIV is designed to learn continuously, keeping track of its inferences in a growing database. Unlike language learning programs such as NELL, the Never Ending Language Learning program developed by computer scientist Tom Mitchell and his group at Carnegie Mellon University, VIV has the advantage of interacting with the real world, including locations, the user's range of interests, and indicators of tastes and preferences. With a wider range of movement and knowledge, VIV can perform calculations correlating trade-offs such as time versus money, quality versus price, and proximity versus distance. It interacts with the user's cognitions in directing movements, interpreting sensory signals such as location indicators, and responding to queries. Moreover, it has the advantages of massive storage space in the cloud, fast processing speed, and computational intensity of data manipulation. VIV and the user, considered as parts of a cognitive assemblage, constitute stable communicating nodes around which hover a cloud of other functionalities associated with web queries and curating algo-rithms that correlate VIV's data with other information about the user stored in myriad relational databases. If the rollout of VIV is success-ful, we can anticipate that it will have considerable commercial value, because it will integrate geolocation with specific product requests. Moreover, it will enable data collection on a scale that surpasses even that available with present smartphones and desktop searches. It will be able to correlate user movements in real time, location data about present trajectories as well as past trips, and the queries and purchases associated with these trips. Some of this functionality is already avail-able in GPS devices, for example the store locations displayed under search categories—information that corporations pay to have listed. When several routes are available, the GPS typically chooses the one that passes by the greatest number of listed stores, for example malls or shopping plazas. All this, and more, will be potentially available with VIV.

Having such a smart digital assistant will also affect how users inte-grate this technical cognition into their daily lives. We can expect that certain evolved cognitive abilities present in human brains—for exam-ple, the ability to navigate and judge where one is in the world—will re-ceive less stimulation with this device, for now it navigates for the user, and the human synaptic networks involved with navigation will tend to shrink. We know from experiments evaluating web scanning versus print reading that human neurology can undergo changes after even minimal exposure to digital media, with long-lasting effects (Hayles 2012). The predictable result is that the more one uses a digital assis-tant such as VIV, the more one needs to use it, as one's natural abilities to navigate in the world decline, perhaps to the point of atrophy.

Moreover, the device will likely result in a certain homogenization of behavior; exploratory wandering will decrease and routinization of movement will increase. The same will be true of shopping practices; less wandering around the store and more direct selection of desired items, where "desire" is itself manipulated by marketing. Overall, the interpolation of the user into corporate designs will be intensified and expanded. To the extent that Augmented Reality may also be part of VIV's functionality, this intensification will occur on nonconscious as well as conscious cognitive levels. As with other digital affordances, VIV will follow what Bernard Stiegler (2010a, 2010b) has characterized as a pharmacological dynamic of poison and cure, offering the powerful advantages of convenience, satisfaction of desires, and enhanced navigation while increasing surveillance, directed marketing, and capitalist exploitation.

Are devices such as VIV the predecessors of fully conscious technical systems, as Spike Jonze's film *Her* (2012) suggests? If the twentieth and twenty-first centuries have taught us anything, it is that only fools would rush in to proclaim "Impossible!" The real temptation here, however, is to imagine that we are *awaiting* the arrival of technical cognition, when it is already realized in myriad complex systems and computational devices. Humans living in societies with deep technological infrastructures are enmeshed in cognitive assemblages shaped and interpenetrated by technical cognition, including language learning systems. If "the language instinct" separates humans from other biological organisms in a rupture so extreme it marks human uniqueness, as Steven Pinker has controversially argued (2007), the language gap is narrowing between humans and their digital interlocutors.

SOCIAL SIGNALING AND SOMATIC SURVEILLANCE

At MIT Media Lab, Pentland, working with his graduate students, has developed a device he calls the sociometer, designed to detect, measure, and display physiological indicators of social signaling among groups (Pentland 2008). The device, worn like a shoulder badge, detects who is talking to whom (via IR transceivers), for how long, and with what intensity (Choudbury and Pentland 2004; Pentland 2008, 99–111). It also analyzes nonlinguistic speech features such as consistency of emphasis, tracks body movements and infers from them the activities involved, measures the proximity to other people, and from these data identifies the social context. Pentland calls the behaviors measured by the sociometer "honest signals" because they take place below the level of conscious awareness; he further argues that attempting to fake them would require so much effort that it is virtually impossible (Pentland 2008, 2–3).

The sociometer performs in a technical mode operations similar to the human cognitive nonconscious by sensing and processing somatic information to create integrated representations of body states. As we have seen, the human cognitive nonconscious recognizes and interprets patterns of behavior, including social signals emanating from others. The importance of this function becomes apparent when it is externalized in the sociometer, for the social signals it detects enable Pentland and his group to predict outcomes for various kinds of interactions, from salary negotiations to dating preferences. Even with thin slices of behavior as short as thirty seconds, sociometer data indicate accurately what decisions a group will reach. It is worth emphasizing that these predictions are derived solely from the sociometer's analysis of social signals, with no attention to verbal content or rational argument.

There are several implications to be drawn from these experiments. First, they indicate the usefulness of the sociometer as a feedback device to help a group improve its functioning. Pentland reports that his lab has developed "a computer algorithm that builds on the sociometer's ability to read the group's honest signaling. Using this technology we are beginning to build real-time meeting management tools that help keep groups on track, by providing them with feedback to help avoid problems like groupthink and polarization" (49).

More fundamentally, sociometer data indicate how essential social signals are to human sociality, and conversely, how more limited rational discussion and conscious deliberation may be than traditionally supposed. In this respect, humans may have something in common with social insects such as bees and ants (as E. O. Wilson argues in another context; Wilson 2014, 19–21). "Honest signals are used to communicate, control the discovery and integration of information, and make decisions" (83), Pentland writes. Extrapolating from this insight, he argues that "important parts of our intelligence exist as *network properties,* not individual properties, and important parts of our personal cognitive processes are guided by the network via unconscious and automatic processes such as signaling and imitation" (88). With classic understatement, he predicts that "we will come to realize that we bear little resemblance to the idealized, rational beings imagined by Enlightenment philosophers" (88).

In addition to this major conclusion, there are important corollaries. Because social signals take time to generate, transmit, receive, and recognize, they operate on a slower timeline than either the cognitive nonconscious (in the 200-ms range) or consciousness (in the 500-ms range). Pentland estimates they work in the 30-second range, a temporality analogous to the time it takes to interpret complex verbal information such as narratives (107–111). This implies that the processing of verbal information and social signals occurs along similar timelines, opening the possibility of a feedback loop between them, such as happens when two people engaged in conversation begin to mimic one another's gestures as they become more engaged with each other verbally, each form of communication reinforcing the other. Pentland references research showing that when this kind of mimicry occurs, interactors report that they like the other more, trust each other more, and reach more accommodating outcomes (10–11). Another corollary is that social signaling affects both participants, unlike verbal language, which can be directed at another without necessarily affecting the speaker. Pentland summarizes this effect: "When you engage in social signaling, you are often affected just as much as the other person. Signals are two-way, not one-way, so that pulling on one corner of the social fabric stretches all members of the network" (40).

Evolutionarily, it is likely that social signaling developed before language; many mammals use such signals to negotiate over territory, communicate intentions, and coordinate group activity. Pentland references brain research indicating that "we all have *networking hardware* that gives us the ability to read and respond to other people" (37), specifically, the mirror neurons mentioned in chapter 2 (Barsalou 2008; Ramachandran 2012). "This signal-response channel seems to have evolved much earlier than the linguistic channel," he continues, "with language building on top of the capabilities of this channel" (42), a trajectory analogous to nonconscious cognition developing first, with consciousness emerging later and being built on top.

The sociometer, then, may be regarded as an exteriorization of the cognitive nonconscious, collecting, interpreting, analyzing, and displaying information in ways that make it available for conscious consideration. Insofar as social signals are essential to effective group functioning, we may draw a somewhat different conclusion from those that concern Pentland. The sociometer's exteriorization reveals that the work of the cognitive nonconscious is crucial to social networking, group decision making, and indeed human sociality. Functioning as part of a cognitive assemblage, the sociometer is enmeshed in a human-technical system with porous borders, depending on who is using the device for what purpose—whether to monitor one's own responses, to surveil those of someone else, or to analyze group behavior from the outside, so to speak, analyzing its dynamics with or without the group's permission .

The significance of this opening-out suggests the need for a new category of analysis that I call somatic surveillance. While traditional surveillance technologies focus on exterior appearances, movements, clothing, and such, somatic surveillance registers the body's interior markers and displays them in real time. The concept is not entirely new. Lie detector tests, for example, measure physiological responses such as heart rate and galvanic skin response and display them through a moving graph; hospital medical devices display heart rate on monitors. Although these technologies operate in real time, they lack two crucial features illustrated by the sociometer—mobility and miniaturization, properties necessary to make the technology wearable (for an analysis from a medical viewpoint of the importance of mobility, see Epstein 2014). The idea of using wearables for somatic surveillance is still relatively new, and its implications remain largely unexamined.

This state of affairs catalyzed a Dutch researcher, Frans van der Helm, to create what he calls the MeMachine. Internationally renowned for his cutting-edge research in robotics and prosthetics, van der Helm and his lab have developed hardware devices and software programs to aid them in their research, including an animated display of an anatomical figure mirroring in real time the movements of a subject wearing multiple tracking sensors on his body. Because their work requires detailed knowledge of how muscles, tendons, and joints move, the figure is imaged sans skin to reveal more clearly the complex interactions that constitute human mobility. This figure, appearing somewhat like one of the flayed bodies in Gunther von Hagens's *Körperwelten* (*Body Worlds*), is all the more dramatic because it moves in perfect synchrony with its real-life counterpart. The capability to create this image had already been developed for research purposes, so it required only a few tweaks to make it into the occasion for a spectacular demonstration of somatic surveillance: the MeMachine.

In 2012, van der Helm demonstrated the MeMachine in a videotaped presentation before a large audience. The video shows the preparations: an anonymous hairy chest being shaved so sensors can be attached; sensors being carefully placed on each finger; a headband with eye-tracking devices slipped into place; a backpack with the computer equipment communicating wirelessly with the sensors hefted onto shoulders. Preparations complete, a nattily suited van der Helm strides on stage wearing this gear, his face partially obscured by the headband, gloves concealing the finger sensors, clothing the ones on his body. On screen, however, *everything* is revealed: as the flayed figure mirrors his posture, gestures, and body language, data scrolls underneath showing ECG readings of his heartbeat, EMG (electromyography) tracking of electrical impulses generated by his muscles, galvanic response indicating skin conductance, EEG readouts showing his brainwaves, and an inset using the eye-tracking data to show where he is looking. Each data cluster is estimated in real time, so that the audience knows his focus of attention, arousal level, emotional state, brain activity, and muscle tension. The kinds of information that the cognitive nonconscious processes are here opened out to an unprecedented degree. Moreover, in Stelarc-like fashion, he suggests in his videotaped lecture that in future instantiations of the MeMachine the audience may be invited to intervene in his somatic states, for example, by setting limits for physiological parameters which, if exceeded, "punish" him by removing some of his functionality. In this case, the boundaries of the MeMachine's cognitive assemblage would include any number of spectators as well as van der Helm and the technical devices monitoring his activities. The prospect suggests that the MeMachine's exteriorization could create surveillance scenarios in which virtually nothing of the surveilled person's thoughts, affects, and autonomic responses would remain private or capable of resisting exterior scrutiny.

When I spoke to van der Helm after his presentation at Delft Technical University (van der Helm 2014), he framed the project as a provocation designed to catalyze discussions about the ethical, social, and cultural implications of somatic surveillance. He reported that ethicists, philosophers, and others who had seen the MeMachine demonstration were shocked by its surveillance potential, proposing that issues of privacy, public policy, and ethical guidelines should be explored before the technology was developed further. To this view, van der Helm was vehemently unsympathetic, asking why these people had not come to his lab to engage in dialogue and what gave them the right to impose constraints after the fact.[2]

I found his response illuminating; it shows the problems with approaches in which humanists stand aloof from technological projects and deliver judgments on them from an exterior perspective. Suppose, instead, that a humanist had found her way to his lab, attended the weekly lab meetings, asked questions, engaged in discussions, and suggested readings for the group to consider.[3] The final results might then have been very different. Rather than demonstrating the MeMachine as a technologically neutral project (which is how it was presented at Delft), van der Helm might have set it up as a cautionary tale, an occasion for a panel discussion, or as an indication of the technology's dangerous potential, along with a discussion of how safeguards and limitations could be built into the technology. In this sense, we might regard the MeMachine as a missed opportunity for productive collaboration between the sciences and humanities.

In another sense, the MeMachine demonstrates the extent to which the workings of the cognitive nonconscious can be externalized through technical mediation, creating situations in which the human cognitive nonconscious, technical cognition, and human consciousness interact in real time through multiple feedback loops and recursive circular causalities to create a cognitive assemblage of unprecedented surveillance potential.

DISTRIBUTED AGENCY AND TECHNICAL AUTONOMY

Cognitive technologies show a clear trajectory toward greater agency and autonomy. In some instances, this is because they are performing actions outside the realm of human possibility, as when high-frequency trading algorithms conduct trades in five milliseconds or less, something no human could do. In other cases, the intent is to lessen the load on the most limited resource, human attention, for example with self-driving cars. Perhaps the most controversial examples of technical autonomy are autonomous drones and robots with lethal capacity, now in development. In part because these technologies unsettle many traditional assumptions, they have been sites for intense debate, both within the military community and in general discussions. They can therefore serve as test cases for the implications of distributed agency and, more broadly, for the ways in which cognitive assemblages interact with complex human systems to create new kinds of possibilities, challenges, and dangers. To limit my inquiry, I will focus on autonomous drones, but many of the same problems attend the creation of robot warfighters, as well as nonmilitary technologies such as self-driving cars, and quasi-military technologies such as face-recognition systems.

The present moment is especially auspicious for analyzing technical autonomy, because the necessary technical advances are clearly possible, but the technical infrastructures are not so deeply embedded in everyday life that other paths are "locked out" and made much more difficult to pursue. In short, now is the time of decision. Debates entered into and choices made now will have extensive implications for the kinds of cognitive assemblages we develop or resist, and consequently for the kinds of future we fashion for ourselves and other cognitive entities with whom we share the planet.

My focus will not be on drone assassinations carried out by the United States in other countries without respect for national boundaries, including some American citizens killed without trial or jury, in clear violation of the Constitution and civil rights. This aspect is well covered by Medea Benjamin's *Drone Warfare: Killing by Remote Control* (2013), in which she passionately opposes the drone program both for its unconstitutionality and more specifically for the horrific toll in civilian deaths ("collateral damage"), estimated to be as high as 30 percent.[4] I also will not consider the myriad uses emerging for civilian UAVs, including rangeland monitoring, search and rescue missions, emergency responders in the case of fire and other life-threatening events, and UAVs used as mobile gateways, or "data mules," collecting data from remote environmental sensors scattered over large territories (Heimfarth 2014). Rather, I will focus on piloted and autonomous UAVs,[5] as well as multivehicle systems proceeding autonomously, with the swarm itself deciding which individual will play what role in an orchestrated attack. This range of examples, showing different levels of sensing abilities, cognitions, and decisional powers, illustrates why greater technical cognition might be enticing and the kinds of social, political, and ethical problems it poses.

With the massive shift after 9/11 from state-on-state violence to what Norman Friedman, a military analyst, calls expeditionary warfare, targets are not associated with a geographically defined entity but with highly mobile and flexible insurgents and "terrorists." Friedman points out that if surveillance can be carried out without the enemy's ability to perceive it, then the enemy is forced to devote resources to hiding and concealing its positions, which not only drains their ability to carry out offensives but also makes it more difficult for them to organize and extend their reach. These factors, he argues, combine to make UAVs superior to manned aircraft for expeditionary warfare. A fighter jet can typically stay aloft only for two hours before it needs to refuel and the pilot, fatigued from high altitudes, needs to rest. In contrast, the UAV Global Hawk can stay aloft literally for days, refueling in the air; with no pilot aboard, pilot fatigue is not a problem (PBS 2013). These factors have led to a significant redistribution of resources in the US Air force, with more UAV pilots currently being trained than pilots for all types of manned aircraft combined. Spending about $6 billion annually on drone development and purchase (Human Rights Watch 2012, 6), they currently deploy about 7,000 drones, compared to 10,000 manned aerial vehicles (Vasko 2013, 84).

The factors that have made UAVs the contemporary weapons of choice for the US Air Force required the coming together of many technological advances, including global satellite positioning, superior navigation tools, better aerodynamics for increased stability and fuel economy, increased computational power, and better sensors for visual reconnaissance, obstacle avoidance, and ground coordination. The weak link in this chain is the necessity to maintain communications between the UAV and the remote pilot. As long as the UAV's performance requires this link, it is subject to disruption either internally, as when the remote pilot banks too suddenly and the link is lost, or because the link has been hijacked and control wrested away by another party, as happened when a Lockheed Martin RQ170 Sentinel drone landed in Iranian territory in December 2007, likely because the UAV was fed false GPS coordinates by the Iranians. This implies that the next wave of development will be UAAVs, unmanned vehicles that fly autonomously, and UAAVS, multivehicle autonomous systems. Still at a nascent stage, UAAVs and UAAVS are nevertheless developing rapidly. The Navy, for example, is developing the experimental X-47B Stealth UAAV, which can perform missions autonomously and land on an aircraft carrier without a remote pilot steering it. Moreover, the technical components necessary to make UAAVs and UAAVS reliable and robust are coming together very quickly in transnational research projects, particularly in the United States and China.

A recent article written in English by Chinese researchers illustrates the growing awareness of UAAVS of their internal states as well as the environment (Han, Wang, and Yi 2013, 2014). The study discusses the development of software that allows a swarm to coordinate its individuals in cases where one or more vehicles are assigned to attack. The model uses an "auction" strategy, whereby each unit responds to a request for a bid by assessing what the authors call its "beliefs," "desires," and "intentions," which are calculated with weighted formulae resulting in a quantitative number for the bid. The software enables the swarm to balance competing priorities in rapidly changing conditions, taking into account their position, velocity, and proximity to one another ("beliefs"), their assigned mission priorities ("intentions"), and the intensity with which they will execute the mission ("desires"), with the latter parameters tailored for specific mission objectives. The anthropomorphic language is not merely an idiosyncratic choice, for it indicates that as the sensory knowledge of external and internal states, autonomous agency, and cognitive capabilities of the swarm increase, so too does their ability to make decisions traditionally reserved for humans.

With autonomous drones and other autonomous weapons on the horizon, there has been increased attention to the ethical implications of their use. Most of these discussions refer to the Geneva Conventions and similar protocols, which require that weapons must "distinguish between the civilian population and combatants" (Article 48 of Additional Protocol 1 to the Geneva Conventions, cited in Human Rights Watch 2013, 24). In addition, international humanitarian law prohibits disproportionate attacks, defined as ones that "may be expected to cause incidental loss of civilian life, injury to civilians, damage to civilian objects, or a combination thereof, which would be excessive in relation to the concrete and direct military advantage anticipated" (from the *Customary International Humanitarian Law Database,* cited in Human Rights Watch, 24). Finally, additional requirements are the rather vague concept of "military necessity," defined by British scholar Armin Krishnan as dictating that "military force should only be used against the enemy to the extent necessary for winning the war" (2009, 91), and the even vaguer "Martens Clause," intended to cover instances not explicit in the Geneva Conventions. It requires that weapons be consistent with the "principles of humanity" and the "dictates of public conscience" (Human Rights Watch, 25).

In assessing autonomous weapons in these terms, Human Rights Watch and the International Human Rights Clinic (IHRC) at the Harvard Law School argue that autonomous weapons, including autonomous drones, cannot possibly make the required distinctions between combatants and civilians, particularly in the context of insurgent tactics that deliberately seek cover within civilian populations. Peter Singer, for example, instances the case of a gunman who shot at a US Ranger with "an AK-47 that was propped between the legs of two kneeling women, while four children sat on the shooter's back" (Singer 2010,

303). They also argue that "proportionality" and "military necessity" are violated by drones, although "necessity" clearly is itself a moving target, given that what constitutes it is obviously context dependent and heavily influenced by the kinds of weaponry available.

The Geneva Conventions were, of course, forged in the aftermath of World War II, characterized by massive state-on-state violence, fire-bombings of cities, gratuitous destruction of cultural monuments, and the nuclear holocausts wreaked by the United States upon Naga-saki and Hiroshima. With the move to expeditionary warfare, rise of insurgent attacks, and continuing increases in US drone attacks, these protocols seem badly outdated, even inappropriate. Why keep coming back to them? On an even deeper level, why should we care about eth-ics in wars where the intent is precisely to kill and maim? Why, as Peter Singer puts it, try to determine what is right when so much is wrong, a question that drives straight to the oxymoron implicit in the phrase "Laws of War" (309). His defense is that the Geneva Conventions, obso-lete as they may be, are the only international accords on the conduct of warfare we have and that we are likely to have, short of another world war with even more devastating violence. He believes there is an important distinction between those who practice restraint in warfare and "barbarians" (309) who are willing to go to any extreme, however savage and brutal. To argue thus is to plunge into a definitional abyss, since what counts as "restraint" and "barbarism" are as contextually and culturally dependent as the distinctions they propose to clarify.

A better approach, I argue, is to evaluate the ethical questions sur-rounding piloted and autonomous drones from the relational and pro-cessual perspectives implicit in the idea of a cognitive assemblage. Mark Coeckelbergh, one of the few philosophers urging a relational perspective on robotics, observes that most ethical theories carry over to robotics the assumption characteristic of liberal individualism, taking "the main object of ethical concern [as] the individual robot" (2011, 245). In contrast, he argues that "both humans and robots must be understood as related to their larger techno-social environment" (245). Regarding ethical issues through the perspective of a cognitive assemblage foregrounds the interpretations, choices, and decisions that technical and human components make as information flows from the UAV's sensors, through choices performed by the UAV soft-ware, to interpretations that the sensor and vehicle pilots give to the transmitted data, on to the decision about whether to launch a missile, which involves the pilots, their tactical commander, and associated lawyers, on up to presidential advisors and staff. Autonomous drones and drone swarms would operate with different distributions of choices, interpretations, and decisions, but they too participate in a complex assemblage involving human and technical cognizers.

The choice, then, is not between human decision versus technical implementation, which is sometimes how the situation is parsed by commentators who prefer a simplistic picture to the more realistic complexities inherent in the situation. As Bruno Latour (2002) argues, changing the means of technical affordances always already affects the terms in which the means are envisioned, so ends and means mutually coconstitute each other in cycles of continuous circular causality. That said, the human designer has a special role to play not easily assigned to technical systems, for she, much more than the technical cognitive systems in which she is enmeshed, is able to envision and evaluate ethical and moral consequences in the context of human sociality and world horizons that are the distinctive contributions of human conscious and nonconscious cognitions. Consequently, we need a framework in which human cognition is recognized for its uniquely valuable potential, without insisting that human cognition is the whole of cognition or that it is unaffected by the technical cognizers that interpenetrate it. Understanding the situation as a cognitive assemblage highlights this reality and foregrounds both the interplay between human and technical cognitions and the asymmetric distribution of ethical responsibility in whatever actions are finally taken.

Although the cognitive assemblage approach can provide useful perspectives on ethical issues, it does not by itself answer the urgent question of whether autonomous drones and drone swarms should be developed by the US military, and if developed, under what circumstances they should be deployed. Arguing in the affirmative, Ronald Arkin, a roboticist at Georgia Tech, envisions an "ethical governor" (2009, 69) that would be built into the weapon's software requiring the weapon first to determine whether the presumed target is a combatant, and then to assess whether the proportionality criteria are met. This proposal strikes me as naïve in the extreme, not only because of the ambiguities involved in these determinations, but more fundamentally, because of the presumption that the weapon's designers would agree to these criteria. Even if the US military decided to do so, when autonomous weapons designed by other states and nonstate

entities fail to incorporate these restraints, would not "military necessity" dictate that the United States do likewise?

The issue, then, cannot be resolved through technical fixes but requires considered debate and reflection, an insight that highlights the
importance of human cognizers as they act within cognitive assemblages. Ultimately the humans are the ones that decide how much
autonomy should be given to the technical actors, always recognizing
that these choices, like everything else within a cognitive assemblage,
are interpenetrated by technical cognition. A chorus of voices argues
that fully autonomous weapons should not be developed and certainly
not deployed. Human Rights Watch and the IHRC, in their thought-
ful white paper considering the issue, conclude, "The development
of autonomous technology should be halted before it reaches the
point where humans fall completely out of the loop" (36). The summary emerging from the Stockdale Center's year-long program on
"Ethics and Emerging Military Technologies," which culminated in
the Tenth Annual McCain Conference on Military Ethics and Leadership, reaches a similar conclusion: "extreme caution should govern the
actual deployment of autonomous strike systems" (Stockdale Center
2010, 429). Even before deployment, however, they write, "We strongly
advise against incorporating 'strong artificial intelligence' in such systems, which would render them capable of learning and even choosing
ends, inasmuch as strong artificial intelligence is highly likely to introduce unpredictability and/or mitigate human responsibility" (429–
30). Noel Sharkey of the University of Sheffield is more blunt; he is
quoted by the website *Defense One* as saying, "Don't make them fully
autonomous. That will proliferate just as quickly and then you are going to be sunk" (Tucker 2014). The risk of proliferation is real; already
fifty-five countries have the capacity to manufacture or build arsenals
of UAVs. Friedman's appendix listing UAVs larger than fifty kilograms
in use around the world runs to a massive 220 pages (Friedman 2010).
Matthew Bolton of Pace University in New York City puts the issue
of lethal force deployed autonomously through UAAVS eloquently.
"Growing autonomy in weapons poses a grave threat to humanitarian
and human rights law, as well as international peace and security. . . .
Death by algorithm represents a violation of a person's inherent right
to life, dignity, and due process" (Tucker 2014).

As these analyses recognize, the possibility of developing autonomous weapons signals a tectonic shift in how warfare is conducted.

The masses of humans that required nations for mobilization, as in World Wars I and II, can potentially be replaced by masses of combat aerial vehicles, all operating autonomously and capable of delivering lethal force. UAVs can now be built for as little as $500–$1,000, making it possible to field a 2,000-vehicle swarm for as little as a million dollars, a sum that emphatically does not require the deep pockets of a national treasury. This enables an entirely different kind of warfare than that carried out by single UAVs, and it brings into view the possibility that UAAVS could be used to launch massive attacks almost anywhere by almost anyone.

This nightmare scenario is worked out in fictional form in Daniel Suarez's *Kill Decision* (2013). With a fair bid to become Tom Clancy's successor, Suarez is like Clancy in including detailed descriptions of military equipment and operations, but he is considerably more critical of US policies and more sympathetic to utopian impulses. *Kill Decision's* plot revolves around Linda McKinney, an entomologist studying the warlike weaver ants in Africa, targeted for assassination by unknown persons and saved at the last moment by a covert black ops force led by the enigmatic Odin. It seems that the shadowy antagonists have marked for assassination anyone with in-depth knowledge of the technologies they are using to create UAAV swarms, including McKinney, because the swarms recognize colony mates and coordinate attacks using the same kind of pheromone signals she found in the ant colonies.

As Odin and his crew (including two ravens) track the UAAVS, it becomes apparent that the drones represent a new kind of warfare. The clarion call comes first in targeted drone assassinations within the United States (an eventuality almost certain to arrive someday in real life). At first their nature is concealed by the government calling them bombs set off by terrorists, and the victims seem randomly chosen with no connecting thread. But a larger plot is afoot, although the perpetrators remain in the shadows. All we know for sure is that the timing is coordinated with a US appropriation bill to create a large fleet of autonomous drones.

The question the text poses, then, is whether autonomous drone warfare on a massive scale is inevitable, given the advantages of UAAVS over manned aircraft and piloted UAVs. Unlike these, autonomous drones have no limit on how far they can be scaled up and thus are able to mass in the hundreds or thousands. In the fiction, the UAAVS are a colony of ship-attacking drones inhabiting a huge commercial freighter, each autonomous but responsive to the colony's chemical signals. The freighter is on course to intercept the US fleet as it carries out military exercises in the South China Sea. The predictable result would be the annihilation of US ships and an international incident blaming the Chinese for the attack. It begins to look as if the drone legislation might be approved, plunging the United States and subsequently the world into a new era of automated drone warfare. Odin observes that "with autonomous drones, you don't need the consent of citizens to use force—you just need money. And there might be no knowing who's behind that money either" (261).

As the fiction suggests, Suarez believes that autonomous weapons must be constrained by international treaty before we are plunged into a *Terminator*-like scenario in which we are no longer able to control the proliferation of these weapons. He also implies that the only political circumstance in which this is likely to happen is if the United States, having exploited its technological advantage in being the first to develop and use drone technology, reaches the stage where drone proliferation by other state and nonstate entities becomes inevitable. In a report issued by the Council on Foreign Relations, Micah Zenko writes, "The current trajectory of U.S. drone strike policies is unsustainable. Without reform from within drones risk becoming an unregulated unaccountable vehicle for states to deploy lethal force with impunity" (2013, 4).

Ironically, the threat of unlimited drone warfare may be the strongest motivation for the United States to reform its own drone policies first, in order to argue for international accords prohibiting further proliferation. The situation is analogous to the United States being the first to develop—and use—nuclear weapons, but then when other states acquired nuclear capability, being a leader in arguing for international controls. However cynically conceived, this strategy did rescue the world from all-out nuclear war. Nuclear weapons, of course, require massive resources to develop and build, which largely limits them to state enterprises. Autonomous drones are much cheaper. Whether the same strategy would work with them remains to be seen.

HUMAN EMOTION AND TECHNICAL COGNITION

So far my argument has emphasized the ways in which human and technical cognitions interact, but their cognitive capacities nevertheless have distinctive differences. On the technical side are speed, computational intensity, and rapid data processing; on the human side are emotion, an encompassing world horizon, and empathic abilities to understand other minds. Roboticist Arkin tends to present human emotion as a liability in a warfighter, clouding judgment and leading to poor decisions (2010). However, emotion and empathy also have positive sides; considered as part of a cognitive assemblage, they can make important contributions.

The French theorist Grégoire Chamayou refers to suicide bombers as "those who have nothing but their bodies" (2015, 86). Applied to groups such as Al-Qaeda and the Islamic State, this is obviously incorrect; they also have AK-47s, rocket grenades, suicide bombs, IEDs, and a host of other weaponry. There have, however, been instances of resistance by those who indeed have nothing but their bodies: the lone student confronting a tank in Tiananmen Square, the hundreds of satyagrahis (resistors) who followed Gandhi to the Dharasana Salt Works in India and were beaten by British soldiers. Intentionally making oneself vulnerable to harm for principled reasons has the capacity to evoke powerful emotions in those who witness it, as world outrage over the episode at the Dharasana Salt Works demonstrated. Vulnerability, whether intentional or not, can also evoke strong emotions in those who perpetrate violence, in some instances leading them to reject violence as a viable solution.

Such is the case of Brandon Bryant, who performed as a drone sensor pilot for the US Air Force for almost six years until he refused to go on, turning down a $109,000 bonus to reenlist. When he finally sought therapy, he was diagnosed with post-traumatic stress disorder. The diagnosis represents, as journalist Matthew Power notes, "a shift from a focusing on the violence that has been done to a person in wartime toward his feelings about what he has done to others—or what he's failed to do for them" (2013, 7). Chamayou sees this shift as a cynical tactic by the US military to claim that drone pilots suffer too, but this interpretation fails to do justice to Bryant's feeling that he has been through a "soul-crushing experience" (Power, 7). Granted that drone pilots suffer far less harm than those they kill or maim, the fact that some of them experience real "moral injury" (Power, 7) can be understood as one of the contributions human emotions make to cognitive assemblages—something unique to biological life-forms that has no real equivalence in technical systems.

Along with emotion, language, human sociality, and somatic responses, technological adaptations are crucial to the formations of modern humans. Whether warfare should be added to the list may be controversial, but the twentieth and twenty-first centuries suggest that it will persist, albeit in modified forms. As the informational networks and feedback loops connecting us and our devices proliferate and deepen, we can no longer afford the illusion that consciousness alone steers our ships. How should we reimagine contemporary cognitive ecologies so that they become life-enhancing rather than aimed toward dysfunctionality and death for humans and nonhumans alike? Recognizing the role played by nonconscious cognitions in human/technical hybrids and conceptualizing them as cognitive assemblages is of course not a complete answer, but it is a necessary component.

For the cultural critic, knowing precisely how the informational exchanges operate within a cognitive assemblage is a crucial starting point from which to launch analyses and arguments for modifications and transformations, deployments or abstentions, forward-moving trajectories or, as a contrary example, international accords banning autonomous weapon systems. Providing the conceptual scaffolding for such analyses is therefore a profoundly political act, self-evidently so in military contexts but also in many other everyday contexts in which technical nonconscious cognitions interpenetrate human systems, such as those instanced in this chapter.

We need to recognize that when we design, implement, and extend technical cognitive systems, we are partially designing ourselves as well as affecting the planetary cognitive ecology: we must take care accordingly. More accurate and encompassing views of how our cognitions enmesh with technical systems and those of other life-forms will enable better designs, humbler perceptions of human roles in cognitive assemblages, and more life-affirming practices as we move toward a future in which we collectively decide to what extent technical autonomy should and will become increasingly intrinsic to human complex systems.

Temporality and Cognitive Assemblages: Finance Capital, Derivatives, and High-Frequency Trading

Whereas chapter 5 illustrated how cognitive assemblages work by surveying sites ranging from urban infrastructure to personal assistants, this chapter focuses on a more delimited set of practices enmeshing human and technical cognition in finance capital, specifically financial derivatives. At issue are not only flows of information, choices, and interpretations, but also the tendency of these cognitive assemblages to create temporal loops that quickly spiral out of control, ungrounded by referential stability.

Another issue is the vast difference between the speeds at which technical cognizers operate in high-frequency trading (HFT), versus the cognitive timelines of humans in the assemblage. Combined with faster processor speeds, vast increases in computer memory, and fiber optic cables through which information travels at near-light speeds, HFT has introduced a temporal gap between human and technical cognition that creates a realm of autonomy for technical agency. Within the space of this "punctuated agency," algorithms draw inferences, analyze contexts, and make decisions in milliseconds. HFT appeared on the scene when derivatives had already begun to dominate financial trading. The complex temporalities inherent in derivatives interact with the changed temporality of HFT to increase further the fragility of financial markets and their vulnerability to feedback loops and self-amplifying dynamics. Analyzing these effects opens a window on how the interpenetration of technical and human cognition functions to redefine the landscape in which human actors move.

The predominance of dueling algorithms has created a machine-machine ecology that has largely displaced the previous mixed ecology of machines and humans, creating regions of technical autonomy that can and do lead to catastrophic failures. In this sense, then, this chapter continues the discussion at the end of chapter 5 about the importance of human cognition in the design process and the essential role it plays in designing the dynamics of cognitive assemblages, especially in specifying the kinds of autonomy technical cognizers will have and the regions within which technical autonomy will operate. In the case of HFT, among the consequences of technical autonomy as it now operates are increased frequency of "black swan" events that threaten the stability of financial markets, a radically changed distribution between human and technical agency, and the emergence of complex ecologies between algorithms that undermine, if not render obsolete, the efficient market hypothesis that underlies many economic models. Because HFT significantly increases risk and exacerbates inequalities, it is an important site to explore how systemic approaches can intervene effectively to create more sustainable, just, and equitable results.

The key to achieving this, as we will see, is altering temporal ecologies so that humans can exercise greater decision powers and machines have more limited scope for autonomous actions. As in the previous discussion about autonomous weapon systems, the point here is that recognizing how cognitive assemblages operate provides crucial resources for constructive intervention and systemic transformation.

COMPLEX TEMPORALITIES AND DERIVATIVES

Following Michel Serres, Bruno Latour developed the idea of temporal folding in technical artifacts. Here is his description, taking as its occasion an ordinary hammer. The hammer, he writes, encapsulates a heterogeneous temporality, including "the mineral from which it has been moulded, the oak which provided the handle . . . the 10 years since it came out of the German factory which produced it for the market" (2002, 249). Not only is the past folded within it, the future is as well, in the sense that the hammer anticipates and helps to call into being the future, for example, the nails it will be used to drive. (Perhaps a clearer example would be a hammer gun, which not only catalyzed the development of nail cartridges but also new construction materials suited to rapid nailing patterns.)

All technical artifacts encapsulate such heterogeneous temporalities, but the temporality of derivatives is especially complex because of their nature as future contracts. To illustrate, we can consider the verb form called in English the future perfect, and more elegantly in French, the futur antérieur, or future anterior. A curious mixture of future and past, the future anterior is represented by the compound verb "will have been," as in "I will have been done by next week." It looks to the future (*will*), then pivots and looks back on this point as if it had already happened (*have been*), in this way stapling the future to the past through an articulation in the present. This doubling-back movement has significant consequences for questions of value. Yesterday's newspaper, had we been able to read it day before yesterday, would be invaluable; today it is worthless. The cliché forms the basis of the film *Paycheck* (2003), where Michael Jennings (Ben Affleck) completes a job for a multinational, the details of which have been wiped from his memory, only to find that his past self has traded a multimillion dollar paycheck for an envelope of seemingly worthless trinkets, which, however, subsequently prove vital to his survival, as his past/future self had already (fore)seen.

As derivative trader and cultural theorist Elie Ayache points out in *The Blank Swan: The End of Probability* (2010), also involved in folding time is a potential or actual change in context, created through the operation of *writing*. As written contracts, derivatives look to the future through their expiry date, and then as it were turn back to see this future date as having already happened to correlate it with a strike price (the derivative's value at expiration). This complex temporality, already in effect when the first derivative was written, has in the contemporary period grown exponentially more powerful and pervasive through the use of HFT algorithms, which create their own version of the future anterior through the microtemporalities in which they operate. These temporal regions are inherently inaccessible to human cognition, which can follow their electronic traces only in a temporal window that has, from the viewpoint of the algorithm, already faded into the past, but which for the inquiring human resides in the future of conscious recognition.

Derivatives may be considered a form of writing, as Ayache suggests, and this writing happens within contexts that are inherently complex because of the fold in time that derivatives enact. Context and writing/reading are bound together in dynamic interplays; each helps to stabilize the other while also operating as flowing mutabilities that transform through time. The situation is delightfully illustrated in Borges's fiction "Pierre Menard, Author of the *Quixote*" (1994), in which Menard, a twentieth-century French critic, is attempting to re-create the *Don Quixote*—not to copy or reconstruct it from memory but to write it afresh out of his own imagination while nevertheless adhering, word for word, to Cervantes's original. As Borges observes, thoughts springing easily to pen for Cervantes have become almost unthinkable for the twentieth-century writer. "Truth, whose mother is history" (1994, 53) is Borges's chosen example; what could this mean for someone whose history includes the holocaust, Orwell's *1984,* and political spinmasters, not to mention deconstruction? Ayache uses Borges's fiction to point out that writing, as durable inscription, has the potential not only to smooth and linearize context by subjecting its unruly chaotic streaming into a uniform flow of language, but also to create a rupture or break that changes context incrementally by inter-acting with it. Arjun Appadurai also emphasizes the written nature of derivatives in his recent book *Banking on Words: The Failure of Language in the Age of Derivative Finance* (2016), in which he argues that the mar-ket crash of 2008 was essentially a failure of language.

The complex temporality inherent in the future anterior is squared for derivatives, for they transform the contexts into which they are in-scribed as any writing does, but beyond that, derivatives change con-texts further through their operations as derivatives. To illustrate, we can briefly review how derivatives are written and exchanged. In its simplest (vanilla) form, a derivative is a kind of insurance or hedge for an underlying asset. Say that you purchase 100 shares of stock A for $100 each, believing it will go up in price. But you also know that it may go down, so you buy a derivative that lets you sell the stock for $100 at a specified future date. If the stock goes up, well and good, you have made a profit. If it goes down, your loss is limited to the derivative price, say $10. The hedge works in the opposite direction if you think the stock will go down; in this case, you purchase a derivative that lets you sell the stock at a future date at a certain strike price, even if the market price by then is lower. Alternatively, you may sell the option without ever buying the shares, in a move that greatly increases your leverage (ratio of assets controlled to investment) over buying the stock itself. This illustrates the process by which derivatives trade in their own right, independent of whether the underlying asset is purchased.

Since the value of a derivative changes over time, its price must be modeled using probabilities. In the famous differential equation developed by Fisher Black and Myron Scholes (1973), with further additions by Robert Merton (called BSM, after its three creators), the derivative price is calculated as a function of the underlying stock price, the riskless interest rate (for example, the interest rate of a certificate of deposit), and the "implied volatility," or the rate of change of the underlying asset's price. Since all the parameters are known except the derivative price and implied volatility, one may assume a value for the implied volatility and solve for price, or conversely run the BSM "backwards" and use the prevailing market price to calculate the implied volatility. The higher the volatility (that is, the more the underlying stock price varies as a function of time), the higher the price of the derivative, for more change implies more risk in hedging the underlying asset.

The great accomplishment of BSM, and the reason its development correlated with an explosive increase in the derivative market, is that it shows how to price derivatives so as to hedge the underlying asset to make risk disappear (at least in theory!). This is accomplished through a strategy known as "dynamic replication," in which the amount of underlying stock held is constantly changed as the stock price (and consequently the derivative price) changes over time. The BSM makes several assumptions at variance with reality (e. g., there are no transaction costs and one can buy or sell at any time without affecting the prevailing price), but undoubtedly the most important is that price variations will follow an equilibrium model. When price variations are graphed, this model assumes that they will follow a normal curve distribution, resulting in the famous bell curve. (Usually the logarithm of the square of the variations is used; this is called the log-normal distribution). At this point Ayache, building on Nassim Taleb's "Black Swan" argument that highly unlikely events may nevertheless have large-scale effects, intervenes to point out that decisive ruptures can and do occur, in effect rendering the equilibrium assumption untenable. "The Black Swan," he writes, is "a non-probability-bearing event of change of context or, in other words, a truly contingent event" (2010, 11).

What then determines the prices of derivatives? According to Ayache, the market itself. Recall that one may solve BSM for the implied volatility if the price is known. Since the market establishes the price, it also determines the volatility, which implies that it operates in a kind of recursive loop grounded on nothing other than the market's own performance. Hence for Ayache, the market is "a matter of ontology, even of a fundamental kind" (2010, 11). The scary (to me) conclusion he asserts is that the market is inherently unpredictable, and the most one can say is that the market is simply whatever happens next. Although the market may seem to follow an equilibrium model for some periods of time, it is always open to contingent and unforeseeable developments. Writing derivatives, which Ayache characterizes as "contingent claims," injects into the present an untamable and unforeseeable future, through the fold in time intrinsic to their operations. It is this aspect that Warren Buffet highlighted when he famously characterized derivatives as financial weapons of mass destruction (Buffet 2002).

The fragility derivatives introduce has been exacerbated by the huge expansion of over-the-counter (OTC) derivatives, which remain entirely unregulated. When Brooksley Born, head of the Commodity Futures Trading Commission from 1996–99, sought to expand the agency's regulatory power to OTCs, she met with huge resistance and, eventually, legislative action prohibiting the CFTC from regulating them, detailed in PBS's documentary "The Warning" (2010). As the documentary's title suggests, her concerns in retrospect seem entirely justified. Shortly after her attempt to regulate OTCs, Long-Term Capital Management (LTCM) in effect went bankrupt, a failure in which derivatives played a major role, as detailed below.

The future anterior temporal loop, brought into existence by writing derivatives, makes derivative writing a form of extrapolation cut free from its underlying grounding base, free to float where the winds of chance blow it. As Brian Rotman (1987) puts it in his analysis of derivatives based on currency exchange, "Any particular future state of money when it arrives will not be something 'objective,' a reference waiting out there, determined by 'real' trade forces, but *will have been* brought into being by the very money-making activity determined to predict its value. The strategies provided by options and futures for speculation and insurance against money loss caused by volatility of exchange interest rates become an inexplicable part of what determines those rates" (1987, 96, emphasis added). In using the future anterior verb form, Rotman retraces in his analysis the folding in time that derivatives perform. Time folded, twice.

TRAUMA, REPRESSION, AND THE MARKET

But surely this cannot be correct, you may object. Is not Ayache's ascription of perfect contingency contradicted by the very equilibrium behavior that markets typically follow? Indeed, Ayache's vision of the market's ontological power is a neoliberal fantasy run wild, fueled by Quentin Meillassoux's (2010) philosophical argument for the absolute nature of contingency and applied by Ayache to finance capital. The flip side of this coin is what Scott Patterson (2010), in his breezy, tabloid-inflected (infected?) account of the tremendous fortunes and enormous egos of hedge fund traders, calls The Truth, the belief that the market is rational, obeys consistent laws, and follows a predictable trajectory. In view of the pervasive search for The Truth through over a century of trading, it is worthwhile to look more closely into the assumptions behind an equilibrium model.

Bill Mauer (2002), in his article on the "theological repressed," starts his account of the history of equilibrium models with Abraham de Moivre, who in 1773 discovered that measurement errors consistently followed what we now call the bell curve, the distribution that consistently appears when random events are charted. For de Moivre, the bell curve was evidence of divine design, God's thumbprint that reassured man there was a hidden order even in seemingly chance events. As the world became more secular, Mauer argues, the explicit acknowledgment of divine design in normal distributions dropped away, or more precisely, became deeply embedded as an unacknowledged presupposition, which he calls the "theological unconscious." He points to the fact, for example, that when college students are asked to do a series of measurements and their results vary from the bell curve distribution, they often fudge them and thus reenact, albeit unknowingly, the theological unconscious. In the same way that Freud argued the unconscious is founded on an original trauma that is subsequently repressed, only to have it return as a symptom, so the theological unconscious, Mauer argues, is formed by repressing the theological association that de Moivre thought he discerned, only to have it return as a symptom, in this case behaviors that blur the line between reality and the model of reality, the stock market as it actually behaves and the equilibrium models that purport to describe it. This is precisely Donald MacKenzie's point (2008) when he calls a financial model "an engine, not a camera," arguing that models drive the market rather than merely reflect its preexisting reality.

With BSM, arguably the market's most influential model, equilibrium assumptions enter through Merton's idea that stock prices follow a "random walk" similar to the random (Brownian) motion of molecules, resulting in the normal curve described above. This assumption is closely tied with the efficient-market hypothesis (EMH), or the idea that markets are "informationally efficient," so that all participants have the same information about past and current developments, and that this information is already taken into account in the prices. In other words, EMH assumes that prices accurately reflect the state of the world as it exists at any given moment. As a corollary, the model implies that in the long run, one cannot make excess profits over what the index averages achieve (a result confirmed by empirical data). BSM, in modeling what the "optimal" prices for derivatives should be, paradoxically also enables traders to identify derivatives that are mispriced, either selling above or below what the BSM model indicates, offering an opportunity for arbitrage (the simultaneous purchase and sale of an asset to benefit from a price difference). The model's predictions here are taken as the *ideal* standard from which the market prices may deviate, another case of a self-referential loop that becomes insulated against its mismatch with reality by operations that tend to bring reality in line with the model.

MacKenzie (2008), in his careful analysis of how well the BSM has matched historical derivative prices, divides the history into "three distinct phases" (202). The first period is from before April 1973 when the Chicago Board Options Exchange opened and the first year of its operation, when there were "substantial differences between patterns of market prices and Black-Scholes values" (202). The second period goes from 1976 to summer 1987, when the prices matched Black-Scholes well. Then the third phase goes from 1987 to 2005–06 (the time MacKenzie was writing), when the fit has been poor, "especially for index options" (that is, options tied to the major exchange indexes such as Standard and Poor).

MacKenzie's explanation for this is fascinating. He conjectures that once traders began using BSM, their activities brought about a better fit between actual prices and the model's predictions in the manner noted above, because they were using the model to decide when derivatives were under- or overpriced and acting accordingly. The tsunami that occurred in the market crash of October 1987, MacKenzie suggests, was so traumatic for traders who lived through it that it left a permanent mark on their psyches, causing them to price options higher than the model indicates, as a kind of unconscious compensation for their perception of higher market risk. In Mauer's terms, we may see this as a symptom not of the theological unconscious but of its opposite: the affective (and not merely conscious) response to finding out that God's thumbprint does not after all mark every trade; sometimes wild gyrations result in crashes that seem more devilish than divine.

The lingering trauma of the 1987 crash, MacKenzie suggests, results in the volatility smile, volatility smirk, or volatility skew, so-called because the volatility plotted against strike price, instead of being flat as BSM predicts, is curved upward or sideways. "The skew seems more extreme than can be accounted for simply by the extent to which empirical price distributions depart from the log-normal assumption of the Black-Scholes-Merton model," MacKenzie writes (2008, 204); he continues, "prices may have incorporated the *fear* that the 1987 crash would be repeated" (205, emphasis in original). Lest this explanation seem far-fetched, we should remember the generation that lived through the Great Depression of 1929–39 and the ways in which it marked them for life. I remember my grandmother, whose husband died abruptly in spring 1929, leaving her with four young children to support (until then she had never worked outside the home). Later in life she saved scraps of string and used them to crochet shelf-edging, a decorative touch her frugality forbade her unless it could be done with material she would otherwise have thrown away. Of course, that generation has now disappeared into dust. The 1987 crash was 28 years ago, and fresh-cheeked traders who were not born when it happened are today writing derivatives. If the fear lingers, what mechanisms are generating it, and what kind of dynamics create ruptures so fierce as to render equilibrium models worthless?

FEEDBACK LOOPS: THE ACHILLES HEEL OF PROBABILITY

To answer these questions, I will find it useful to refer to the detailed case studies MacKenzie performs to determine the causes of spectacular market crashes. One such study is the failure of Long-Term Capital Management (LTCM), a company trading in derivatives; like the crash of 1987 he similarly investigates, feedback loops played a prominent role. LTCM was started in 1994 by John E. Meriwether, and boasted of having Myron Scholes and Robert Merton on its board of directors, the creators of BSM, who would win the Nobel Prize for their work in 1997 (Fisher Black was deceased by then, ruling him out as a Nobel Prize winner). Essentially, the firm's trading strategies revolved around arbitrage, in the sense that they would look for anomalies or mispricings and create options to take advantage of them. An example MacKenzie cites is the price of 30-year treasury bonds versus the "odd year" twenty-nine-year bonds (2008, 216–17). The 30-year bonds typically sold for more than the 29-year ones, but as the maturity date came closer, the prices tended to converge, so LTCM would create an option that shorted the 30-year bonds and went long on the 29-year bonds. If held to maturity, the option would necessarily make the firm money, but the problem was losses in the meantime that could lead to calls for more collateral, a serious issue because LTCM sometimes had leverages as high as 40:1 (although the average was 27:1, not unusual for a hedge fund).

When Russia announced on August 17, 1998, that it was defaulting on bonds denominated in rubles, the event initiated a "flight to quality," with investors fleeing illiquid or risky ventures for more liquid or safer ones. According to MacKenzie, LTCM had anticipated that some event might lead to a flight to quality, and for this reason required their investors to leave their capital in the firm for three years (2008, 230). Nevertheless, the relative prices (in the example above, between 30-year bonds and 29-year bonds) began to widen, and as the spreads widened, arbitrageurs reduced their positions, in effect making the price pressures even more intense.

As the losses mounted, LTCM was taking huge hits, losing as much as 44 percent of its capital in August 1998. The death blow was the monthly newsletter Meriwether sent out to his investors. Although he took the position that this was an extraordinary buying opportunity, investors when they saw the numbers withdrew their capital in droves, forcing LTCM into crisis, teetering on the brink of failure. Indeed, Meriwether was correct; if LTCM could have held out long enough, it would eventually have been able to profit handsomely from the precipitous drop in prices. The situation recalls John Maynard Keynes's well-known aphorism, "The market can remain irrational longer than you can remain solvent" (qtd. in Lowenstein 2001, 123). Time had run out. On September 20, officials from the New York Fed and Assistant Secretary of the Treasury Gary Gensler met with the board of LTCM, and quickly brokered a deal in which fourteen of LTCM's largest creditors agreed to infuse $3.6 billion into the fund, in return for which they received 90 percent ownership.

While admitting that a flight to quality was a major factor, MacKenzie argues that there was another force at work. LTCM's success and extraordinary profits (as high as 40 percent in some years) catalyzed a host of imitators, who tried to infer from LTCM's moves what economic models it was following and adjusted their models accordingly, creating what MacKenzie calls a "superportfolio" (MacKenzie 2005). In effect, LTCM's success invited imitation, and the more imitators, the less diverse the investment landscape became, and consequently more fragile and vulnerable to disruption (a well-known result in ecology). Although the specifics were different than for the 1987 crash (portfolio insurance and mechanical selling in the former case, arbitrage in the later instance), one similar aspect stands out: both crashes involved feedback loops that broke out of the parameters of equilibrium models and created self-reinforcing spirals downward.

Because LTCM had Scholes and Merton on its board, many commentators drew the conclusion that when the firm crashed and burned, it showed the BSM model simply did not work. Others attributed the firm's demise to greed, excessive risk, and/or over-leveraging, but MacKenzie is at pains to emphasize the safeguards that the firm put in place and the (relatively) conservative nature of its estimates and cash reserves. The lesson I draw from his analysis is this: even careful planning, stress testing, and Nobel Prize–winning minds are not sufficient to guard against the consequences of the temporal folding that derivatives enact, with consequent instabilities in value and ruptures of context. When conditions are right for feedback loops to emerge, they will erupt, frequently with devastating consequences. Moreover, history demonstrates that sooner or later, conditions will be right, as they were again in the crash of 2007–8 and the resulting Great Recession, from which we are still recovering.

GLOBALIZATION, "EXORBITANT PRIVILEGE," AND THE 2007–8 FINANCIAL CRISIS

Were feedback loops also a major factor in the crisis of 2007–8? Certainly they played a part, but a full panoply of financial sins was operating as well: overleveraging, excessive risk, duplicity, exploitation, double-dealing, and old-fashioned human greed. Derivatives were involved primarily in two guises: through the use of credit default swaps, and as options written on subprime mortgages. Credit default swaps insure a creditor from default on a loan; the major US institution offering credit default swaps in the run-up to 2007–8 was AIG, American International Group. AIG offered insurance on derivatives written on bundled assets in which subprime mortgages were "tranched," that is, sliced into pieces, divided into different risk classes, and sold as Collateralized Debt Obligations, or CDOs. The idea was that repayments from mortgages would go first to the highest tranche, and only when those obligations were satisfied would the returns flow down to the next tranche, and so forth. The highest-grade tranche was rated by rating agencies (themselves paid by the very institutions they were supposed to be evaluating) as AAA investments, even though the basis for all the tranches were risky investments with higher-than-normal rates of default. Moreover, credit default swaps, like derivatives in general, could be traded in their own right without owning the underlying asset of subprime mortgages, a fact that increased the speculative bubble as more investors poured in.

When large numbers of these subprime mortgages began defaulting as economic conditions tightened, AIG was unable to meet all of its obligations and, facing a liquidity crisis and possible bankruptcy, received a bailout of $85 billion from the United States Federal Reserve Bank, on the grounds that it was "too big to fail"—the largest bailout in history at that point. Eventually this ballooned into $183.2 billion as Treasury provided additional funds, and AIG was subsequently forced to sell off numerous assets to repay its government loans. As credit tightened because of the large number of loans in default, almost every sector of the economy was affected. The feedback loops here were global in scope, plunging the US stock market into free fall and spiraling out to stock markets in China, Europe, and elsewhere.

The story, then, is not only about the US economy but the world financial system. MacKenzie reports a retrospective reflection by Myron Scholes about the failure of LTCM; "maybe the error of Long-Term was . . . that of not recognizing that the world is becoming more and more global over time" (MacKenzie 2008, 242). As our scope widens to encompass the world economy, I turn to Yanis Yaroufakis's analysis of the "exorbitant privilege" (2013, 93) that the United States enjoys by having the dollar as the world's reserve currency.[1] When the Bretton Woods agreement was nullified by President Nixon in 1971, taking the US dollar off the gold standard and allowing currencies to float relative to one another, the United States' hegemonic status and the reputed stability of the dollar attracted investments into the United States, both in terms of Treasury bonds and corporate debt vehicles.

As Yaroufakis points out, the exponential growth of derivatives from the 1970s onward in effect served as a huge increase in the amount of private money. Since derivatives are contracts written in the present against some future event, they leverage underlying assets such as stocks, related to the "real economy" through ownership in corporations, by creating financial instruments that have their own market and exchange values. The more those markets increase, the more the financial economy expands, even though the "real economy," as Robert Brenner (2006) has forcefully argued, may have been stagnating and even declining since the 1970s.

Coupled with this inflationary money supply are the twin deficits that the United States has been running since the 1970s, the national debt and the trade deficit. Yaroufakis asks rhetorically, "And who would pay for the red ink? Simple: the rest of the world! How? By means of a permanent tsunami of capital that rushes ceaselessly across the two great oceans to finance America's twin deficits" (2013, 22), flowing from such surplus countries as China and Germany. Writing with soul-cleansing anger and white-hot rhetoric,[2] Yaroufakis traces the transition from the post–World War II Global Plan, when America used its surplus to invest in Europe and Japan, to the Global Minotaur, the image he uses to describe the flows of money from surplus countries to the United States, where consumers purchase the products produced by those same countries, which increases debt, which is further financed by capital abroad and further increases the twin deficits, in a continuous cycle. Yaroufakis argues that the crisis of 2007–8 rendered this cycle untenable. However that may be, it is self-evidently not a sustainable model, since the twin deficits cannot keep increasing forever without consequences. According to Yaroufakis, the demise of the Global Minotaur has left the global economy without a viable mechanism for the flow of international trade and capital, and the Great Recession cannot permanently be put behind us until this problem is solved.

Dick Bryan, associate professor of political economy at the University of Sydney, and coauthor Michael Rafferty, associated with the Faculty of Commerce at the University of Wollongong, Australia, offer another explanation (2006) for the exponential growth of derivatives after the 1970s. They argue that the abstract nature of derivatives provides a way to achieve commensuration over time and space, thus serving to anchor the transnational flows of money and capital after the gold standard was abandoned. The point can be illustrated with the CDOs composed of subprime mortgage slices (my example, not theirs). Any one mortgage is tied to a particular house serving as collateral and a particular debtor responsible for making the mortgage payments. If the debtor defaults, the creditor is out of luck, and so banks lending money for mortgages have traditionally been careful to make sure the debtor is creditworthy and can afford the loan. However, if that same property is tossed in with hundreds of others and sold as a CDO, the situation changes significantly. First, the firm selling the CDOs does not suffer if they become toxic, for they have made their commission in the sale. Second, the creditors buying the CDOs insure them through AIG, so if they go bad, AIG is responsible for making them whole. When AIG itself teeters on bankruptcy, the government intervenes with a bailout that makes AIG whole, dollar for dollar. Thus the risk is transferred again and again until it ultimately winds up with the taxpayer, who foots the bill. Add in the commensuration effect, and the risk is now spread worldwide.

There are two consequences to this system. The first is that derivatives, functioning as a form of private money as Yaroufakis argues, keep inflating the money supply and pumping more and more money into these cycles. The second is that the financial economy now comprises an ever-growing percentage of total economic activity, while the real economy languishes. The result is to make it more crucial than ever to analyze the effects of financial derivatives, including how HFT algorithms and technical cognition are transforming the landscape of finance capital. This leads us back to cognitive assemblages and another dimension of the complex temporalities derivatives enact, this time in the incommensurable timelines of human and technical cognizers.

HFT ALGORITHMS AND THE FLASH CRASH OF MAY 2010

An example will illustrate how trading algorithms engage in continuous reciprocal causality through recursive feedback loops, sometimes with results that defy human reason. A third-party book on Amazon, entitled *The Making of a Fly,* shot up in ten days from an initial price of $199 to $24 million—to be precise, $23,698,655.93, plus shipping (reported on CNN April 25, 2011; see http://www.michaeleisen.org/blog/?p =358 for an account). How did this absurd price emerge? One seller's algorithm priced the book at 1.27 times the price of another seller's algorithm, which would in turn revise its price to 0.998 times the price of the first algorithm. For example, if the book was listed at $100 by the second seller, the first seller's algorithm would price it at $127. Reacting to this revised price, the second algorithm would price it at $126.75. The first algorithm then prices the book at $160.96, and so on, up to millions of dollars. Of course a human, with a deeper knowledge of the world and wider world horizon, would have known the figure was ridiculous.

When and why did the market revolution happen that moved the action from human traders to automated trading algorithms and HFT trading in particular?[3] *Wall Street Journal* writer Scott Patterson traces the history of automated trading (2012). He recounts its origins in the programs Joshua Levine designed for Datek, a firm that traded on Instinet, a private exchange created for firms to buy and sell stocks directly to one another, avoiding the fees charged by the NYSE. When Datek asked for lower fees from Instinet and was turned down, the company went out on its own and Levine created Island, a new kind of electronic pool in which algorithms, retrospectively discovered and given names like Dagger, Sniper, Raider, and Stealth, fought each other, using cutting-edge artificial intelligences to sell and buy with reaction times measured in milliseconds. Levine sent out an e-mail to users of his Watcher program, writing in January 1996, "We want Island to be good and fair and cheap and fast. We care. We are nice. SelectNet is run by Nasdaq. They don't care. Instinet is run by Reuters. They aren't nice . . . Won't you join us at Island" (qtd. in Patterson 2012, 121).

When Levine wrote that Instinet was not "nice," he was referring to the ways in which HFT traders ripped off other traders and indeed their own clients, helped by exchanges such as Instinet that cloaked their bids from public view. A trader could, for example, offer to buy a large block of stock for a client, which would drive the price higher, while offering to sell the same stock over Instinet at the new higher price. The client would not know about this side deal, because he could not see the Instinet side of the offer. In other practices, traders would front run a stock. For example, if a trader submitted an order to purchase 100,000 shares of Company X, an HFT with a faster algorithm would detect the bid, purchase the stock and relist at a higher price, a practice called "sniffing," about which we will hear more shortly.

The complexities of trying to regulate a large-scale technical system driven by the capitalist imperative to maximize profits is difficult even in the best of times. A case in point is the SEC's attempt to make the market fairer by introducing Order-Handling Rules that created new entities called Electronic Communication Networks, or ECNs, and imposing certain restrictions that forced markets such as Instinet to list their quotes on Nasdaq. Patterson summarizes the effects of ECNs: "With the Order-Handling Rules, the entire Nasdaq marketplace would shift toward an electronic platform wide open to computer-driven trading . . . the phone-based system of human dealers would quickly become a screen-based cyberpunk network of computer jockeys born and bred in electronic pools such as Island" (2012, 128).

With the floodgates open to computerized trading, speed became the name of the game. Stock trading companies are willing to pay high fees to have their computers located on rack space within the server farms run by the major exchanges (a practice called colocation), because the proximity shaves milliseconds off the time it takes to transfer information from the exchange to the trader, time that can be turned into money by taking advantage of the small price differences that occurred in that time interval. Currently an ultrafast transatlantic cable is being constructed, at the cost of several billion dollars, to connect high-frequency traders in the United States with those in Britain. The estimated time acceleration is five milliseconds, which works out to about a billion dollars per millisecond advantage (Johnson et al. 2012, 4).

As a result of HFT, the average time a stock is held has plummeted across all the exchanges. After World War II, it clocked in at four years; by the turn of the millennium, it had dropped to eight months; by 2011, it had, according to some estimates, dived to an astonishing twenty-two seconds (Patterson 2012, 46). The amount of information being exchanged on stock trades correspondingly grew to gargantuan figures. Nanex, a firm that tracks speed trading, estimates that on all US stocks, options, futures, and indexes on a single day, one *trillion* bytes of data are exchanged (Patterson 2012, 63).

As speed trading increased, the duopoly of NYSE and Nasdaq shattered into a number of private markets such as Instinet as well as "dark pools," trading sites where quotes are hidden from public view. Attempting to cope with this fragmentation, the SEC introduced in 2007 a new set of regulations called the National Market System, or Reg NMS. The idea was to bind together all the electronic marketplaces into a single network so they would operate as a true national market. At the heart of the Reg NMS was the mandate that any order to buy or sell a stock must be routed to the market that had the best price. If there was a disparity, for example, between the price a stock sold on Nasdaq and a higher price listed on the NYSE, the order would automatically be routed to the Nasdaq. To facilitate this mandate, an electronic ticker tape was instituted called the Security Information Processor, or SIP feed.

One consequence of Reg NMS was that now the trading houses had to monitor the prices in all venues, all the time, which virtually forced them to use cutting-edge algorithms. Moreover, there were also unintended consequences. Orders are executed according to their place in the "queue," the electronic monitoring system that assigns them a priority according to the time they were placed. But, as Patterson explains, "an order in one exchange queue could be suddenly rejected, routed to another exchange, or kicked to the back of the queue if an order that beat its price" appeared (2012, 49). This provided a host of new ways to game the system. High-frequency traders had by this time become the largest customers of the exchanges; by 2009, it is estimated that they accounted for 75 percent of all trades. In addition, the NYSE had transformed from a nonprofit entity to a for-profit corporation when in April 2005 it merged with a private exchange, Archipelago (modeled on Levine's Island), and then a couple of years later with Euronext, the European combined stock market. For its part, Nasdaq had changed in 2006 from a quotation service to a licensed national exchange, and by 2011, more than two-thirds of its revenue came from HFT (Lewis 2014, 163).

Operating as for-profit corporations with their own stockholders to please, the exchanges instituted new order types that went far beyond the old-fashioned limit orders (orders to buy or sell within specified price limits) that were the bread and butter of exchanges in previous decades. Moreover, competition had become so fierce that high-frequency traders were making less on each deal, and as a consequence they became more dependent on "maker and taker" fees, a structure the exchanges had put into place to maximize liquidity.

To ensure liquidity, a trader who provides it is typically given a small rebate, while a trader who takes it has to pay a small fee (called the "maker/taker" policy). Patterson summarizes the effects when the new order types were combined with the rebate/fee structure. The new type of orders "allowed high-frequency traders to post orders that remained hidden at a specific price point at the front of the queue when the market was moving, while at the same time pushing other traders back. Even as the market ticked up and down, the order wouldn't move . . . By standing at the front of the queue and hidden as the market shifted, the firm could place orders that, time and again, were paid the [rebate or 'make'] fee. Other traders had *no way of knowing* that the orders were there. Over and over again, their orders stepped on the hidden trades, which acted effectively as an invisible trap that made other firms pay the 'take' fee" (Patterson 2012, 50).

This situation makes clear the emergence of a new type of hyper-capitalism that I call vampiric capitalism. As Mark Neocleous observes (2003), Marx in *Capital* uses the vampire as a metaphor for capital sucking the lifeblood from the working class; by contrast, vampiric capitalism preys on other capitalistic enterprises. These practices illustrate how far the stock market had drifted from its initial purposes. As Sal Arnuk and Joseph Saluzzi (2012) explain, the stock market was originally set up to enable new companies to attract capital by issuing IPOs, thus spurring innovation and creating diversity in the marketplace. It also provided a way for ordinary people with disposable income to invest in the stock market through mutual funds, options, and other investment instruments. High-frequency traders perform none of these useful services. They justify their existence by claiming they provide liquidity for the marketplace (Perez 2011, 163), and because they trade so often, they have also driven down the spread between buy and sell orders, which they argue benefits everyone. But as we shall see, there is another side to this story. Although their commissions are smaller, because they trade so often, the pennies quickly mount into dollars and, eventually, into billions a year—money that ultimately comes out of the pockets of investors. The deleterious effects on innovation may be seen in the alarming decrease in publicly traded companies: in 1997, there were 8,200 public companies; by 2010, only 4,000 remained (Patterson 2012, 59).

The most disastrous effects, however, are the instabilities that HFT introduces. Nanex, a company that analyzes the algorithms high-frequency traders use, detected an algorithm it called the Disruptor.

The Disruptor is designed to flood the market with so many orders that, effectively, it disrupts the market itself (Patterson 2012, 63). These instabilities became shockingly evident on May 6, 2010, when in the space of two minutes the Dow fell 700 points, and then just as quickly rebounded. How did this happen? In an already down market nervous about possible defaults by Greece and Spain, Waddell & Reed Financial from Kansas City was monitoring a massive order to sell seventy-five thousand S&P 500 E-mini futures contracts, worth about $4 billion dollars. The algorithm they were using was designed to sell at a pace that would keep it at about 9 percent of the market's overall volume, with thirty-second pauses to throw off the shark algorithms hunting for "whales" (large orders) so they could front run them. These sell orders were bought by high-frequency funds that, within milliseconds, sold them again to other high-frequency traders, sometimes at a slightly lower price; these algorithms in turn sold them again. As the volume shot up and the market plunged down, the trading algorithm from Waddell & Reed increased its selling because its 9 percent limit kept increasing, which prompted an even fiercer feedback cycle. Patterson reports that "within a fourteen second period high-frequency traders bought and sold an astonishing twenty-seven thousand E-mini contracts" (Patterson 2012, 264).

As the market plunged, other stocks were affected as well. Accenture, a global consulting company that normally sold for about $50 a share, dropped to an absurd price of a penny a share, a so-called "stub quote" that trading firms place simply to fulfill their obligations as market makers but that they never expected to have filled. Procter & Gamble plunged from its normal price of $60 a share to around $30. On the other end of the scale, some stocks, notably Apple, demonstrated an amazing upward spike, reaching $99,999 per share (no doubt another stub quote). As the insanity proceeded, many high-frequency traders, alarmed by the volatility and concerned that transactions would be retroactively cancelled by the SEC, simply pulled the plugs on their computers. This meant that even fewer buyers were in the market, accelerating the market's downward plunge. Only when NYSE shut down trading for five seconds was the feedback cycle broken, and at that point, because prices on many stocks were ridiculously low, the algorithms started buying and, within a couple of minutes, the market was back to where it was before. But not before inflicting real harm on some buyers. One person who was trying to sell Procter & Gamble just before its price bottomed out lost $17,000, and a hedge fund in Dallas lost several million dollars when the price of options it was buying spiked from 90 cents to $30 per contract.

Subsequently, the SEC revoked or "broke" trades that exceeded a 60 percent variation from their prices before the flash crash. Then the SEC, in coordination with the Commodity Futures Trading Commission (CFTC), set about to investigate the flash crash, issuing a report in September 2010 (while the flash crash took only five minutes to plunge and recover, they took a full four months to figure out what happened). The report focused on liquidity, concluding that Waddell & Reed's sell order was the culprit, initiating a chain of events that "sucked liquidity out of the market and allowed prices to go into freefall" (Buchanan 2011). Mark Buchanan, a theoretical physicist who blogs on financial matters, references the research done by Eric Scott Hunsader, founder of Nanex and a software engineer. Hunsader followed 6,483 trades that Waddell & Reed made that fateful day; Buchanan notes that "the company's execution broker fed the market throughout the day—a tactic specifically designed to minimize the price impact of a large sale" (Buchanan 2011, 2). Instead of locating the problem in this sale, Hunsader's analysis "suggests this plunge was caused by high-frequency traders. They typically act as liquidity providers, standing ready to buy and sell at certain price levels. But the day's volatility prompted them to dump their holdings to avoid losses . . . It was this selling, not Waddell & Reed's passive orders, that caused the liquidity to disappear" (Buchanan 2011, 2).

The comments to Buchanan's article are revealing. "Guest" called the SEC report a "fairy tale," and "Fritz Juhnke" commented that "it is about time someone pointed out that 'making trades' is not the equivalent to providing liquidity. The exchanges have hoodwinked the regulators into believing that it is acceptable to pay brokers for making trades. The farce is that the exchanges call it 'providing liquidity.' Regulators need to wake up to the fact that sales which trigger due to a price decrease are removing liquidity (i.e., exaggerating price moves) rather than providing it" (Buchanan 2011, 3). "SofaCall" commented, "This is why I have been out of the market for the past six years. It's no longer possible to handicap value. All you can do is try to catch the trend—which is not much different from casino gambling. Only the gaming commissions do a better job of assuring that casinos are honest than the SEC does with the financial markets" (Buchanan 2011, 4).

Calling the SEC report "a trend in finance industry public relations strategy," "Matthew" noted that most people "won't understand they are being ripped off if the con is cloaked in a modest level of complexity" (Buchanan 2011, 5). "H_H_Holmes" succinctly summed up the consensus: "The façade that the industry has anything to do with 'people' investing in 'businesses' is gone. All algorithms, all the time. Joe Six-Pak, meet Skynet" (Buchanan 2011, 4).

The significance of the flash crash was underscored by Nanex's research into other miniflash crashes that occurred from 2006–10 but went largely unnoticed because they involved single stocks and unfolded too quickly to attract attention. Flash crashes tend to disappear in statistics using day-by-day figures; a finer-grained temporal metric, such as the one used by Nanex, is necessary to spot them. Nanex's analysis found 254 minicrashes in 2006, before Reg NMS was initiated, and in 2007, when Reg NMS was being rolled out over the course of a few months, 2,576. In 2008, when Reg NMS was fully in effect, that number increased to 4,065. Clearly, the unintended effect of Reg NMS was dramatically to increase the likelihood of flash crashes (www .nanex.net/FlashCrashEquities/FlashCrashAnalysis_Equities.htmlco).

Neil Johnson, a physicist at the University of Miami, published with his colleagues "Financial Black Swans Driven by Ultrafast Machine Ecology" (Johnson et al. 2012), which confirmed Nanex's research. Their abstract notes, "We provide empirical evidence for, and an accompanying theory of, an abrupt system-wide transition from a mixed human-machine phase to a new all-machine phase characterized by frequent black swan events with ultrafast durations." Analyzing 18,530 black swan events between 2006–11, they construct a model that indicates black swan events are more likely as the time duration of trades decreases and as the diversity of strategies diminishes. They assume that the algorithms, having to act so fast, must have only a few strategies from which to choose, and moreover that many algorithms will have very similar strategies, exacerbating the possibilities for feedback loops. The upshot is that black swan events are not anomalous; rather, they reflect the dynamics of a machine-driven trading in which humans play no part in the actual transactions. Reflecting back on the SEC report about the May 6, 2010, flash crash, we can now see that its conclusion that the Waddell & Reed transaction was the culprit is incorrect. It may have been the initiating event (i.e., the last straw), but it is the system dynamics that make such crashes inevitable.

COMPLEX ECOLOGIES OF
HUMAN-ALGORITHM INTERACTIONS

In a study based on sixty-seven interviews with traders in a firm that engaged in high-frequency trading (HFT), Ann-Christina Lange (2015) creates a vivid picture of how humans interact with the algorithms operating on their behalf. She observes, "I was quite surprised to learn that [the traders] talked about their algorithms not as purely rational actors making markets more efficient but rather as interacting agents operating in the market" (1). She records one trader pointing to the screen and saying "'I can see how this guy is moving. I learn from him . . . There is no way I can really know who is behind the algorithms, but I recognize this pattern'" (1). Another trader observes, "'You read other algorithms. They are all based on rules. You come out with a generalized robust set of rules to deal with others'" (5). An analogy is a video game player who infers the rules governing the algorithm generating the screen display by observing the game's behavior when he tries various tactics; with practice, he is able to predict how the game will react in specific circumstances and refines his tactics accordingly, a phenomenon that James Ash calls an "envelope": "Envelopes emerge from the relationship between the user's body and . . . an 'interface environment'" (2016, 9). As Johnson noted above, HFT algorithms typically sacrifice multiplicity of inputs for speed; they usually have only a few strategies and thus are able to be "read" by traders watching their moves.

Moreover, as Lange points out, the algorithms are constantly interacting with other algorithms, generating a complex ecology that, Lange suggests, can be understood as swarm behavior. In human terms, their interactions resemble the kinds of moves and countermoves typical of propaganda (psyops) warfare: feints, dodges, misinformation, and camouflage. MacKenzie (2011), for example, notes that order execution algorithms typically do not place large orders all at once but slice the orders into smaller pieces and feed them into the market gradually, just as the Waddell & Reed algorithm did when trying to execute a large sell order. One kind of algorithm does this through volume-weighted average price, or VWAP (veewap), parceling out the order in proportion to volumes of shares traded in equivalent time slots on previous days. Other algorithms hunt for these VWAP "whales" (large orders), detecting them through "sniffing" and front running their orders to take advantage of this knowledge. In other practices of questionable legality, "spoofing" algorithms place orders to entice other algorithms to respond, and when they do, instantly cancel the orders and use the knowledge to make profits. One trader that Lange interviewed explains, "We want to have as many quotes as possible but also to make sure we don't need to get fills . . . we hope not to have to execute them as they usually don't end up making us money or even lose us money. But the information they provide is very important, where we can know very quickly if something is going to happen. And we don't have to wait for the data feeds update to come in, which is too slow" (Lange 2015). Another of her interlocutors sums up the situation: "it's algorithms fighting other algorithms. You can game the kill switches. Push this algorithm into this position and you know it will pull back and you can profit from that" (Lange 2015).

As Arnuk and Saluzzi have pointed out, the profits generated by HFT are purely speculative and contribute nothing to the real economy other than making money for the trading firms that employ them. As the percentage of trades executed by HFT algorithms continues to increase, the result is that the market ecology moves more and more into speculation and less into synergistic interactions with the real economy.

Not everyone agrees with Johnson's and Nanex's claim that HFT is rendering the market more fragile and unstable. MacKenzie (2011), for example, asserts that the data is not clear one way or the other, although he sounds a note of caution about the instability of complex technological systems such as finance trading has become. Nanex's discovery that Ultrafast Extreme Events (UEEs) are not anomalies but frequent occurrences can be read in two ways: either they do not much matter because they are so small and disappear so fast, or they are like tiny fractures in a space capsule, harbingers of a major catastrophe. In my view, MacKenzie is too cautious in his assessment; the small cracks to my mind are signals that the cognitive assemblages formed by HFT and human actors are systemically risky and are weighted too much toward technical rather than human cognition.

Whatever one makes of UEEs, some effects are indisputable. One, indicated above, is the huge growth of speculative activity through HFT over investing in the real economy. Another is the shredding of the efficient market hypothesis, which readers will recall is the idea that all participants in the market have essentially the same information at the same time, along with the corollary that prices reflect the real state of the world at any moment. In HFT, the algorithms are designed precisely to disrupt equal access to market information and to create informational inequalities by ferreting out information from rival algorithms while concealing their own actions.

A third effect, and perhaps the most important, is the complete transformation of the temporal regimes within which trading occurs, and a consequent "arms speed race" toward faster and faster algorithms, faster and faster connection cables, and faster and faster exchange infrastructures. MacKenzie (2011) writes that Turquoise, the trading platform of the London Stock Exchange, can now execute trades in 129 *microseconds* (that is, a little more than a tenth of a millisecond). The transformation from the world of the pits, where traders stood next to one another in all their sweating, noisy, pushy macrophysicality, has given way to the world of screens and algorithms, endlessly searching for the small perturbations and algorithmic interactions that can make money in the world of derivatives and HFT. Andy Clark (1989, 62), in discussing why consciousness emerged through evolutionary processes, has called it a weapon in the cognitive arms race (61). HFT may similarly be regarded as an evolutionary milieu in which speed, rather than consciousness, has become a weapon in the nonconscious cognitive arms race—a weapon that threatens to proceed along an autonomous trajectory in a temporal regime inaccessible to direct conscious intervention. Any solution, then, must address the cognitive dynamics of the entire cognitive assemblage.

SYSTEMIC REENGINEERING: INVESTORS EXCHANGE (IEX) AND BATCH AUCTIONS

In *Flash Boys: A Wall Street Revolt* (2014), Michael Lewis documents the experiences of Brad Katsuyama, initially a trader for Royal Bank of Canada (RBC) Capital Markets. He notes Katsuyama's mystification when he put in a bid (for example, to buy a stock) and watched the price move away from him as soon as the price was entered. With hindsight, we can surmise what was happening; an algorithm had detected his bid and moved the price higher a fraction of a second later. Katsuyama then embarked on a journey to discover what was happening, querying colleagues throughout the industry to find out how the algorithms were constructed and what they did. According to Lewis, none of the other traders actually knew; here the story does not quite ring true, for it is more likely that at least some of them knew but were unwilling to divulge the information. In any event, when Katsuyama finally discovered what was happening, he felt strongly that it was a perversion of the legitimate purposes that the stock market is supposed to serve (a conclusion Arnuk and Saluzzi [2012] also endorse, as we have seen, and Lewis cites them for their contestations of HFT practices). Unlike almost everyone else who was going along with the game, Katsuyama began a quest to rectify the situation.

He determined to attack the root of the problem: the systemic dynamics that had resulted in a cognitive arms race. Older ideas about how the market operates, including the efficient market hypothesis and regulatory interventions, were rapidly becoming obsolete. Lewis quotes Katsuyama's reaction when he analyzed the SEC report about the flash crash of May 2010: what leaped out to him was its "old-fashioned sense of time." Katsuyama concluded, "'Once you get a sense of the speed . . . you realize that explanations like this . . . are not right'" (Lewis 2014, 81).

Realizing that HFT had effectively rigged the markets so they no longer functioned as unbiased intermediaries, he and his colleagues determined to create a new exchange, which they called simply Investors Exchange (IEX), designed to reinstate the legitimate purposes of stock and derivative trading. The idea was not to speed trades up but on the contrary to slow them down, so that everyone's bid arrived simultaneously, no matter how fast their algorithms were. In addition to this general idea, Katsuyama and his colleagues spent long hours anticipating how the system might be gamed, making it as resistant as possible to predatory strategies.

IEX's battle for credibility and official sanction as an exchange has been a long and arduous uphill fight. Lewis documents that even when IEX was functioning—a small miracle in itself—and investors were instructing their brokers that they wanted their trades routed through it, some brokers were deliberately sidestepping these instructions and sending them to other exchanges instead. After several delays, IEX has won the fight to become an official exchange and not just a trading site.

In the meantime, the IEX website emphasizes what may be described as their *ethical* approach to trading. "Dedicated to institutionalizing fairness in the markets, IEX provides a more balanced marketplace via simplified market structure design and cutting-edge technology," the website reads: http://www.iextrading.com/about/. A short video at the site begins with the question, "What is the purpose of the stock market?" and then proceeds to give a lesson on HFT and predatory trading. It ends with aligning IEX's principles with the Securities Exchange Act of 1934: "uphold just and equitable principles of trade," "remove impediments to a fair and open market," "protect the investors and public interest," and "facilitate an opportunity for investor orders to meet directly." This approach merits serious attention, for it suggests that ethical responsibility is not only possible within a capitalistic system, but that large investors such as institutions handling retirement and pension funds would *prefer* a fair and equitable exchange in which predatory algorithms are not given free rein, for the simple reason that the profits created by HFT in the end come out of their pockets.

A different solution is proposed by Budish, Cramton, and Shim in their article on frequent batch auctions (2015). They document some of the costs of the speed arms race, including Spread Networks's cable from New York to Chicago that cost $300 million and shaved off three milliseconds from the communication time. They also test the claim of high-frequency traders that HFT increases liquidity and measure it against the cost of "sniping," that is, "stale" bids that lag behind market data and are picked off by high- frequency traders before the offering firm has a chance to withdraw them. These costs are figured into the bid-ask spread and therefore increase the cost of trading for everyone. Moreover, in the present system of continuous limit order book (CLOB) trading, in which trades are executed throughout the day, there is a high probability that bids will be sniped because orders are executed serially. Whereas only one firm is trying to retract its stale bid, many high-frequency traders are trying to benefit from it, and the probability is that one of them will succeed in having its order filled before the firm's order can be executed in the queue.

There are other problems with CLOB, as the authors note. They investigated two securities that normally closely track each other, the SPDR S&P 500 exchange-traded fund (SPY) and the S&P 500 E-mini futures contract (ES). Because both are based on the S&P index, one would expect that they would move in tandem with a high correlation, and this is what the data show when the metric is a day, an hour, and even a minute. However, when the temporal metric shrinks to

250 milliseconds, large gaps appear between the movements of the two securities, and the correlation breaks down almost completely. "This correlation breakdown in turn leads to obvious mechanical arbitrage opportunities," they write, "available to whomever is fastest" (2015, 2). Moreover, the breakdown does not lead to an increase in market efficiency, only to an increased arms race in which the window of profitability drops dramatically, "from a median of 97 ms in 2005 to a median of 7 ms in 2012" (2015, 2). The amount of money siphoned off by the arms race is significant. With this pair of securities alone, the authors estimate that annual cost is $75 million. Although they decline to speculate on what the cost would be for all traded securities, saying only that "the sums are substantial," it does not take a rocket scientist to see that the cost would soar to many billions and perhaps even trillions—money that does nothing for the real economy and only serves to enrich HFT firms committed to the arms race.

When the authors modified their initial model, which assumed that all traders had simultaneous access to the same information about prices, to take into account differentials between firms that invested heavily in the arms race and consequently had faster algorithms that those that did not, what emerged was a classic prisoner's dilemma. "All firms would be better off if they could collectively commit not to invest in speed, but it is in each firm's private interest to invest" (2015, 5).

The solution to these interrelated problems, the authors suggest, is to move from CLOB, which treats time as a continuum, to a system of batch auctions executed frequently, say every tenth of a second (100ms). In the batch auction, all bids present at that time would compete against one another, in effect moving the competition from time to price. Moreover, the lack of correlation between securities that in longer temporal metrics move together would be mitigated, again moving the competition from time to price. Because the length of time algorithms had to act would be increased, there would be less incentive to make algorithms simple so they could be fast, encouraging the development of "smarter" algorithms with more diverse trading strategies, thus increasing the depth and diversity of strategies in the algorithmic ecology, with a consequent increase in robustness and a decrease in fragility. Finally, the competition based on price would move the market toward efficiency rather than toward the increasing trading costs associated with the arms race, including sniping, building faster connections through expensive infrastructure, and the resources devoted to building ever- faster algorithms.

These two different solutions—IEX making trading slower, and the batch auction idea—share in common the belief that the answer does not lie in regulation but rather than in systemic (re)engineering. They assume that the profit motive will remain intact, but by changing the ways in which profits can be realized, they move the entire system toward efficiency, fairness, and the productive use of capital maximized for larger social goals and not merely for personal gain by individual trading firms investing in the arms race. Note, however, that in the process, fundamental ideas about how the world is structured become transformed: for the batch auction proposal, time is changed from a continuum to a set of discrete intervals, and for IEX, time is slowed down so that trading occurs closer to the range of human perception and away from computational speed. These are not inconsequential differences. Transformative within the realm of finance capital, they can also potentially spread beyond financial exchanges into other areas of social and cultural life. They illustrate how the interpenetrations of complex human systems with cognitive technical systems form larger techno-epistemo-ontological structures that determine our sense of how the world works. They also show that human agency, deployed with a strong sense of ethical values such as fairness and social good, can create new systemic structures more conducive to human flourishing than can the profit motive unrestrained by other considerations.

FINANCE CAPITAL AND THE HUMANITIES

For better and worse, finance capital is so deeply enmeshed with the self-organizing ecology of ultrafast machine algorithms that it has become impossible to think of our global economy without them. As we have seen, the machine ecology is extraordinarily sensitive to small changes in its environment (i.e., the regulatory framework within which it operates). Every new regulation introduces new ways to game the system, a fact confirmed by historical research (Lewis 2014, 101). Consequently, Katsuyama and his collaborators concluded, "There was zero chance that the problem would be solved by financial regulation" (Lewis 2014, 101). Their solution of attacking the problem through systemic dynamics and design, a strategy also at the center of the batch

auction proposal, offers important clues to the role that the humanities can play in intervening in the world of HFT and finance capital
more generally.

Catalyzed by an idealistic desire to make the markets fairer, Katsuyama and his colleagues put everything on the line to make their
vision a reality. Their commitment illustrates that in a large sense,
the situation is not strictly a technological problem but one of value
and meaning (recall Heidegger's famous argument that the essence
of technology is nothing technological). One of the few humanists
to tackle the issue is Bernard Stiegler. Through a series of important
texts—*Technics and Time, 1* and *Time and Technics, 2* (1998, 2008), *Taking
Care of Youth and the Generations* (2010), and especially *For a New Critique of Political Economy* (2010)—he undertakes the ambitious project
of proposing a general framework through which to understand the
coevolution of humans and technics. Because of its scope, the project
has both impressive virtues and limitations that may not be immediately apparent. The test case proposed here, HFT algorithms, can
help to refine Stiegler's framework and also reveal new insights as to
where the coevolution between humans and technics is heading in the
contemporary period, especially in the evolution of cognitive assemblages, the complex temporalities they instantiate, and the resulting
incommensurable timelines of human and technical cognition.

In *Technics 1* and *2,* Stiegler develops the concept of tertiary
retention—memories stored in artifacts that allow access to events a
person never experienced first-hand. Moreover, he argues that tertiary
retention precedes individual cognition. In *Taking Care,* this formulation is linked with the development of "long circuits" that allow tran-
sindividuation to occur, resulting in an individual becoming intellectually mature and taking responsibility. Subverting the development
of "long circuits" are what Stiegler calls the programming industries
such as television, video games, and the Web, which seek to capture
the attention of young people and convert the tradition of what I have
called deep attention into hyper attention (Hayles 2012).

The case of HFT programs allows us to see in Stiegler's concept of
tertiary retention an unrecognized print-centric bias. While tertiary
retention works well for books as external storage devices, it covers
over the rupture that occurred when technics ceased to be only about
storage and instead became about machine agency and systemic machine ecologies. A book can be said to possess agency when a human

reads it and the act of reading causes the human's cognitive system to work in a different way—a dynamic central to the humanities, with their deep tradition of ideas conveyed through writing or what Stiegler calls a "grammatological" process. Note, however, that this agency is converted from a passive possibility into an actuality only because a human is involved in writing and reading. In contrast, the point of HFT programs is precisely that, once created and set in motion, they do not require any human agency to act. Indeed, humans are deliberately cut out of the circuit to allow the machines access to the microtemporalities essential for HFT. Stiegler is correct in noting that the creation of a "grammatological" process such as alphabetization breaks a continuous flow of words into discrete units such as letters, phonemes, etc. However, using this same term for digitization glosses over the difference between alphabetization and executable program code: while letters are passive, code executed by a machine can actually make things happen without human intervention.

"Long circuits," as Stiegler uses them, apply to human cognition. But machines have "long circuits" too, and the effects here are quite different from those Stiegler attributes to humans. When automated trading programs execute trades, they often do so in circuitous routes in order to optimize certain parameters, for example keeping the fees that brokers pay as low as possible. Their effects, in other words, protect and advance vampiric capitalism, as pointed out by Arnuk and Saluzzi as they diagram the convoluted "wandering path from order-generation to execution" (2012, 146).

The "long circuits" in intelligent machines also relate to anticipation. This too becomes a machine function as the algorithms search for patterns that will enable them to predict the next micromovement of stock prices, as well as orders that competing algorithms may be about to execute. In a sense, these algorithms have both too much memory—hypermnesia—as they race through real-time data streams analyzing what all the other algorithms are doing, and too little memory—hypomnesia—as they jettison the information about what happened yesterday (never mind years or centuries ago) to cope with the huge amounts of information streaming into their processing units. Unlike humans, who must sleep to process their memories, these machines have no unused computer cycles. Even when the markets are closed, they are still gathering data, analyzing it, and placing orders that affect what the opening stock prices will be.

Clearly, then, focusing on memory functions alone is insufficient to understand the complex dynamics of the machine ecologies of automated trading. Also crucial is the agency that the machines possess and their abilities to analyze and predict at lightning speeds, and also to evolve and learn as they compete with other algorithms. Automated trading systems embody evolutionary dynamics that can lead to unpredictable consequences and emergent behaviors. Humans may set up these systems, but they are not in complete control of how they operate, evolve, and mutate. The issue is not memory alone, but a transformation of global economic systems that increasingly drive us toward vampiric capitalism and away from social responsibility.

Another aspect of HFT that has engaged humanists is the transformation of the temporal regimes into microtemporalities inaccessible to humans. A theorist at the forefront of this discussion is Mark B. N. Hansen in *Feed-Forward: On the Future of Twenty-First Century Media* (2015). Working through the processual philosophy of Whitehead, Hansen focuses on what he calls "atmospheric media" designed to fly under the radar of consciousness and influence actions, behaviors, affects, and attitudes before consciousness, with its relatively slow uptake and limited information processing ability, has an opportunity to evaluate the media input. I am not sure if HFT qualify as "atmospheric media" in Hansen's sense, because they are aimed not at affective states but at executing trades within machine-machine ecologies. Nevertheless, HFT certainly use similar microtemporal strategies; they also suggest how the kind of analysis that Hansen enacts might be expanded beyond the affective realm that concerns him.

This point may be developed by comparing Hansen's approach with that of Luciana Parisi and Steve Goodman (2011) in their essay on mnemonic control. Also working from Whitehead, although with a different interpretation, Parisi and Goodman write about the potential of digital media to address the affective body through what Whitehead called prehensions; their essay focuses particularly on branding. When the phenomenon of a prehension goes through further neuronal processes and arrives in consciousness, it will already seem familiar even though it has not been consciously experienced before, leading them to speak of the "past not lived, a future not sensed" (2011, 164) that nevertheless makes one susceptible to branding influences. We can see that their idea of mnemonic control works on a timeline that operates before Stiegler's primary retention (conscious experience), and much earlier than his secondary or tertiary retention. It does not depend upon inscription and recovery at a later time through "grammatological" decoding; rather, control comes in the form of sensations that precede consciousness and directly address the body's affective responses, leading to the cultural and media phenomena now called affective capitalism. Mnemonic control thus exploits the half-second of neuronal processing that occurs before consciousness comes online to create susceptibilities and vulnerabilities used to sell products.

Hansen, by contrast, wants to place the intervention that twenty-first century media make into what he, following Whitehead, calls the "vibratory continuum" or the "worldly sensibility"; importantly, these influences occur prior to sensation and perception. Twenty-first century media, as Hansen uses it, does not mean merely media operating in the twenty-first century but rather is an analytical category for a specific kind of media synonymous with "atmospheric" or environmental effects. These media operate so quickly and pervasively that they intervene in the ground from which prehensions arise. They thus *precede* the cognitive timeline while also decisively influencing the kinds of sensations and perceptions that are possible and relevant within a given milieu.

In terms of this project, Hansen's work is a crucial intervention for at least three reasons. First, it foregrounds the issue of temporality in relation to technical and human cognition, recognizing the incommensurability of their cognitive timelines. Second, it proposes that the effective point of intervention must occur on timelines appropriate to technical cognition, placing it prior to the 100-millisecond range where perception and sensation register for humans. Third, it recognizes that such interventions will be effective only if they are systemic, which he aims to do by both invoking and modifying Whitehead's radically processual view of reality.

A limitation of his analysis from this point of view is that he gives almost no examples of media that operate in this way. The media he mentions—the sociometer, sound art, etc.—work through sensation and perception, not prior to them. For an explication of interventions occurring prior to sensation, we may turn to Nigel Thrift's "technological unconscious," discussed in chapter 5 in relation to urban infrastructure (Thrift 2004, 2007, 91). As noted previously, Thrift sees technical infrastructures as generating a host of presuppositions about the way the world is and how it works, for example, the weight-bearing capacities and kinesthetic qualities of concrete, asphalt, and steel compared to dirt, grass, and swampland. "All human activity," Thrift writes, "depends upon an imputed background whose content is rarely questioned: it is there because it is there. It is the surface on which life floats" (2007, 91). Someone raised in a city would have one set of presuppositions, for example, while another person raised in a tribal area of Africa would have quite another. Multiply these presuppositions by a hundred- or thousandfold operating in every area of life, and you arrive at something like the technological unconscious. Its content "is the bending of bodies-with-environments to a specific set of addresses without the benefit of any cognitive inputs" (Thrift 2004, 177). The technological unconscious is "a prepersonal substrate of guaranteed correlations, assured encounters, and therefore unconsidered anticipations" (177). These presuppositions form, as it were, an interface of high dimensionality mediating between the world and the body, the technical milieu and the humans who learn to navigate and manipulate it.

While Hansen's analysis cogently captures the essential ways in which temporality enters this picture, it is not clear in his analysis why media should have pride of place in the technological unconscious compared to skyscrapers, light rail transit, and a host of other technological infrastructures. However, if we note that nearly all these infrastructural architectures have computational components (as argued in chapters 1 and 5), then the focus returns not to media in general, but to computational media specifically. As David Berry (2011) puts it, "the ontology of the computational is increasingly hegemonic in forming the background presupposition for our understanding the world" (Kindle locations 2531–32).

As we have seen, the qualities that make computational media exceptional are *their cognitive capacities and their abilities to interact with humans as actors within cognitive assemblages.* They thus address human cognitive capabilities across the full range of precognitive and cognitive timelines: as presuppositions preceding sensation, as stimuli producing sensations and perceptions, as input through somatic markers into the cognitive nonconscious, and as experiences within the modes of awareness, consciousness, and the unconscious.

Returning now to HFT, we can see how these ideas serve as resources to grasp more fully the implications of the transition to a machine-machine ecology of automated trading. It is not entirely accurate to say that human intentionality and agency have been sidelined in HFT, but certainly they now operate much more visibly and self-evidently though an elaborate chain of technical mediations than ever before was the case. Personification, always lurking when technical systems exhibit agential powers, becomes almost inevitable, as indicated in the quotations above where traders give personal pronouns and genders to the algorithms they design and observe. In terms of the technological unconscious, perhaps the major predisposition to emerge from these complex human-nonhuman technical systems is the awareness that cognitions happen throughout the system in ways that enmesh them together. Human complex systems and cognitive technical systems now interpenetrate one another in cognitive assemblages, unleashing a host of implications and consequences that we are still struggling to grasp and understand. Nevertheless, the examples of IEX and batch auctions show that human intervention is certainly possible when aimed at systemic dynamics, and that such interventions can and do change the cognitive ecologies to make them more sustainable, more affirmative of human flourishing, and more equitable in their operations.

MEANING, INTERPRETATION, AND VALUES

Let us return now to the urgent question of how the humanities might intervene in the increasingly precarious position into which HFT seems to be leading us. Here Stiegler makes a crucial contribution, for he insists that the fundamental questions concerning technics have to do with meaning, interpretation, and values. Hansen's intervention is also a major contribution, for his intention in *Feed-Forward,* as in most of his work, is to provide theoretical frameworks that enable and catalyze constructive political action. As we have seen, humanists are unlikely to contribute to debates about regulatory reform of the stock markets, a task that requires intimate knowledge about the systemic dynamics to ensure that the reforms will not have deleterious unintended consequences, which in any event may be unavoidable. Humanists can contribute, however, to discussions about the larger social purposes that finance capital is intended to serve. As recent events have vividly demonstrated, the profit motive stripped of all other considerations leads to a disastrous spike upward in systemic risk and, consequently, in global economic instability. Humanists can help to put finance capital in historical perspective and connect it with values such as social responsibility, fairness, and economic justice.

Questions of meaning that occupy the humanities, then, are if anything even more important with the rise of the cognitive nonconscious and the growing importance of cognitive assemblages in finance capital. Because this framework enlarges the realms in which meaning and interpretations can be seen to operate (as detailed in chapter 1), it implicitly creates a bridge between the traditional humanities and the kinds of nonconscious cognitions performed in HFT, as well as between the technical cognitions of the algorithms and those of humans who design and implement them.

What other kinds of initiatives will further this goal? There is already a growing body of research that undertakes this task, including historical research on finance capital (Poovey 2008; Baucom 2005; Lynch 1998), ethnographies of Wall Street (Ho 2009; MacKenzie 2008; Lange 2015), STS-inflected studies of finance (Callon 1998; MacKenzie 2008), and analyses of automated cognition (Parisi and Goodman 2011; Thrift 2007). This emerging field, which lacks a universally acknowledged name, might be called "Critical Studies in Finance Capital," a term that links economic practices with the recognition that the world has a stake in these practices, which consequently cannot and should not be considered only in terms of how much profit they generate.

To be taken seriously in this endeavor, humanists will need to learn the vocabulary, mechanisms, and histories of finance capital. While Stiegler's work has much to offer in this regard, it is couched in terminology that an economist would find completely opaque; Hansen's work, although more lucid, is also dense with argumentation and references that the finance community would likely find difficult to negotiate. The work of building bridges between finance capital and the rich critical and philosophical traditions of the humanities requires that humanists learn to write and speak in ways legible to the finance community, for only so can there be a successful transmission of ideas across these fields. The price to gain admission to discussions with economists, business school professors, traders, politicians, and other influential actors is steep, but the potential contributions humanists can make more than justify the investment. If there is no way out of the global financial system, then the way forward may require going more deeply into it. "Critical Studies in Finance Capital" should be a project in which humanists claim their stakes and make their arguments, transforming it even as we are also transformed by it.

Intuition, Cognitive Assemblages, and Politico-Historico Affects: Colson Whitehead's *The Intuitionist*

As previous chapters have suggested, cognitive assemblages are inherently political. Comprised of human-technical interfaces, multiple levels of interpretation with associated choices, and diverse kinds of information flows, they are infused with social-technological-cultural-economic practices that instantiate and negotiate between different kinds of powers, stakeholders, and modes of cognition. Chapters 5 and 6 explored how these negotiations take place in urban infrastructures and finance capital, respectively. In chapter 6, the focus was on the technical cognitions of algorithms, and the possibilities for systemic transformations by changing the kinds of temporalities involved. In contrast, this chapter focuses on affective forces within assemblages, which as we will see go beyond human responses to the postulated responses of technical artifacts. The chapter takes as its tutor text a novel by African American writer Colson Whitehead, *The Intuitionist* (1999), interrogating how the novel creates richly textured affective, embodied, and interpretive contexts. Through these, the novel shows how the systems making up a cognitive assemblage form connections, create linkages between disparate phenomena, facilitate or block information flows between sites, make choices at multiple levels of human and technical cognitions, and morph as the assemblage gains or loses parts and undergoes systemic transformations in its dynamics.

The Intuitionist nominally takes the electromechanical elevator as its focus but soon invokes a larger cognitive assemblage, progressively widening the boundaries to include the first "colored" female inspector, Lila Mae Watson, city politicians, the Department of Elevator Inspectors and associated Guild, competing elevator corporations United and Arbo, the Mafia, two bouncers at a dance hall, and last but scarcely least, "the most famous city in the world" (33), notorious for its verticality and hence its infrastructural dependence on elevators.

The story is set in an era when integration in Northern cities has a tentative foothold in city job placements but blatant racism is still everywhere apparent, from housing patterns and casual bar conversations to such racist entertainments as blackface minstrel shows. Weaving these diverse strands together, *The Intuitionist* foregrounds two distinctively different modes of cognition, represented by the two factors struggling for power in the Guild, Empiricism and Intuitionism. Whereas Empiricism investigates the soundness of elevators using measurable variables and arrives at results that can be verified empirically, Intuitionism relies on intuition, internal visualization, and feelings to arrive at judgments expressed not through measurements but subjective feelings. The Inspectors who use Intuitionism are called by detractors "swamis, voodoo men, juju heads, witch doctors, Harry Houdinis," all terms, the narrator notes, "belonging to the nomenclature of dark exotica, the sinister foreign" (57–58). By contrast, Empiricism, derided by its opponents as "flat-earthers, ol' nuts and bolts, stress freaks" (58), is identified with a rationality infused with white values, practices, and histories. The issue is political in a literal sense, for the election for the Guild chair looms, in which Orville Lever, champion of the Intuitionists, is pitted against Chancre, an Empiricist and the incumbent chair; by tradition, the winner becomes the head of the Department of Elevator Inspectors.

Historically, of course, the empiricism of such English scientists as James Clerk Maxwell (1871) and William Thompson (Lord Kelvin) played central roles in the expansion of the British Empire by inventing more efficient steam engines, resulting in superior naval power and the subsequent subjection of native peoples in India, the Pacific Islands, and elsewhere, a story repeated with tragic regularity over the globe as tribal peoples came in contact with Western technologies. As Chancre proclaims, "Why hold truck with the uppity and newfangled when Empiricism has always been the steering light of reason? Just like it was in our fathers' days, and our fathers' fathers" (27). Empiricism rules the day—except in Whitehead's fiction, where "No one can quite explain why the Intuitionists have a 10% higher accuracy rate than the Empiricists" (58).

Lila Mae Watson, the first female and the second African American to be hired by the Department of Elevator Inspectors, became a convert to Intuitionism during her student days at the Institute for Vertical Transport, when she encountered James Fulton's two books, *Theoretical Elevators One* and *Two*, works that entranced her and launched Intuitionism. Whitehead gives us a sample of intuitionism methodology when Lila Mae inspects the elevators at 125 Walker.

> She's trying to concentrate on the vibrations massaging her back. She can almost see them now. This elevator's vibrations are resolving themselves in her mind as an aqua-blue cone . . . The ascension is a red spike circulating around the blue cone, which doubles in size and wobbles as the elevator starts climbing. You don't pick the shapes and their behavior. Everyone has their own set of genies. Depends on how your brain works. Lila Mae has always had a thing for geometric forms. As the elevator reaches the fifth floor landing, an orange octagon cartwheels into her mind's frame. It hops up and down, incongruous with the annular aggression of the red spike . . . The octagon ricochets into the foreground, famished for attention. She knows what it is" (6).

Thereupon she cites the elevator for a faulty overspeed governor, the orange octagon's technical correlate. In free indirect discourse, the narrator announces repeatedly that when it comes to elevators, "She is never wrong," following it up in the text's final words, "It's her intuition" (253).

It would be difficult to imagine a more direct assertion of nonconscious cognition and its crucial importance for the modes of awareness. Lila Mae's conscious mind is not in control of the shapes presenting themselves for her inspection: "You don't pick the shapes and their behavior." Recognizing that it "depends on how your brain works," the narrative accepts matter-of-factly that the shapes represent meaningful interpretations of incoming data. Tracing the processes involved, we note that the elevator's performance is sensed and analyzed nonconsciously from incoming somatic data (the vibrations on her back) and then forwarded to core consciousness, which renders the data as visual forms (much like the modal simulations discussed by Barsalou [2008]), whereupon higher consciousness correlates the shapes with technical knowledge available as verbal formulations (a faulty governor). Lila Mae's elevator inspections, then, operate as cognitive assemblages, with cognitive functions distributed over the elevator as technical object reporting on its state, Lila Mae's nonconscious cognition interpreting this input, and her conscious mind delivering a technical diagnosis.

That the elevator, an electromechanical machine lacking electronics and sophisticated computational capabilities, can nevertheless function as a cognitive agent requires a radical shift of viewpoint, as Mr. Reed, mastermind of Lever's campaign and himself a convert to Intuitionism, explains to Lila Mae in discussing Fulton's two volumes. "[Fulton's] diary shows that he was working on an elevator [when he died], and that he was constructing it on Intuitionist principles," he tells her. She responds, "At its core, Intuitionism is about communicating with an elevator on a nonmaterial basis. 'Separate the elevator from elevatorness,' right? Seems hard to build something of air out of steel." He ripostes that "they are not as incompatible as you might think." "That's what Volume One hinted and Volume Two tried to express in its ellipses—a renegotiation of our relationship to objects." He clarifies: "If we have decided that elevator studies—nuts and bolts Empiricism—imagined elevators from a human, and therefore inherently alien point of view, wouldn't the next step, after we've adopted the Intuitionist perspective, be to build the right way. . . ." Grasping his point, Lila Mae finishes the thought: "Construct an elevator from the elevator's point of view" (62).

WHY NONCONSCIOUS COGNITION IS NOT ENOUGH

Tempting as it may be to see this as a conceptual breakthrough of the kind new materialists advocate, we can take a clue that this is not the whole story from the fact that Mr. Reed is soon revealed to be as manipulative as Chancre's gang. He and his accomplices at the Intuitionist House are intent on using Lila Mae for their own designs, which include recruiting an African American corporate agent/spy to their plot. The agent introduces himself to Lila Mae as "Natchez," playing on her sense of racial solidarity to elicit her help to find Fulton's visionary design, called simply the "black box" (61), and subtly romancing her, succeeding in tempting her to drop her customary guard enough to become his ally. The plot then, if nothing else, indicates that realigning human with technical cognition will not be enough to catalyze a reconfiguration of cognitive assemblages profound enough to unsettle entrenched racism and launch the city toward a new kind of trajectory.

Throughout the text, elevator technology serves as a metaphor for "racial uplift," an elevation that propelled the city toward verticality and simultaneously opened a wedge for the first tentative moves of integration. Violence still simmers, however; the text alludes to the recent riots, and other historical indicators suggest it is set in the period following the Harlem riots of 1964.[1] Lila Mae's father makes clear to her, before she leaves her Southern home, that "they can turn rabid at any second; this is the true result of gathering integration; the replacement of sure violence with deferred sure violence" (23). Before she leaves her tiny city apartment in the morning, Lila Mae girds herself with her inspector's uniform, her badge, and her guarded face. "She needs the cut of the suit to see herself," the narrator remarks. "The bold angularity of it, the keen lapels—its buttons are the screws keeping her shut" (56). She survives, she fits herself into the marginal spaces available to her, like the converted janitor's closet she lived in while at the Institute. After graduation she does her job, but her reach is limited, and the further she moves from the city's "zero-point" (4), the more entrenched the opposition, the more suspicious and closed down the faces become.

Two related events catapult her out of survival mode, her customary guarded position in which she reacts to events but does not initiate them. Elevator No. 11 in the Fanny Briggs Memorial Building crashes; it does not just fail but fails catastrophically, going into free fall in defiance of the safety features that supposedly make such an accident impossible. Since Lila Mae had inspected that elevator the day before and given it a clean bill of health, suspicion immediately falls on her, and it is her determination to clear her name that moves her out of her narrow domain, first to the Intuitionist House, then to a detective's peripatetic wanderings as she searches for evidence that the elevator was sabotaged. The second event comes on the heels of the first, when "Natchez" reveals to Lila Mae that James Fulton, founder of Intuitionism and her hero, was an African American passing for white. Although Natchez lies in claiming that he is Fulton's nephew and that he is searching for Fulton's plans for the black box to take back for black people what white people have stolen from them, Lila Mae tests the tale of Fulton's passing against her intuition and decides that part of the story is true.

Armed with this knowledge, she returns to Fulton's two volumes. Like Fanny Briggs, the escaped slave who taught herself to read, Lila

Mae also teaches herself to read. She focuses on this passage in volume two: "The race sleeps in this hectic and disordered century. Grim lids that will not open. Anxious retinas flit to and fro beneath them. They are stirred by dreaming. In this dream of uplift, they understand that they are dreaming the contract of the hallowed verticality, and hope to remember the terms on waking. The race never does, and that is our curse." "The human race, she thought formerly . . . But now—who's 'we'?" (186).

Her movement from reaction to action can be traced through her visits to Marie Claire Rogers, the black housekeeper (and perhaps lover) that Fulton insisted on hiring and ensured her future by trading the promise of his papers to the Institute for her right to continue living in his faculty house at the Institute after his death. On her first visit, Lila Mae goes at the behest of the Intuitionist faction, which hopes she will be able to persuade Mrs. Rogers to turn over the missing journals, and perhaps Fulton's plans for the black box that will revolutionize vertical transport technology. She is unsuccessful, and although Mrs. Rogers confides more to her than to any of the others who came seeking the information, Lila Mae leaves without the papers. On her second visit, however, she goes of her own accord, having decided, based on her new ability to read Fulton's texts for what they hint but do not say, that the transcendental claims he wrote in volume one were intended as a prolonged practical joke against the elevator industry, a satiric response to their insistence on the replication of mundane reality. "Well look at you," Mrs. Rogers replies. "Not the same girl who was knocking on my door last week, are you? . . . You've seen some things between now and then, huh?" (236).

Mrs. Rogers relates that she learned Fulton's secret when his much darker-skinned sister came to visit. Shortly thereafter he began work on volume two, having become a believer in the utopian vision that began as a joke. Her house has been trashed, presumably by the same crew that came after Lila Mae when she discovered that "Natchez," actually Raymond Coombs, was employed by the Arbo corporation as an enforcer. Unexpectedly, Mrs. Rogers gives her the remaining journals, appropriately hidden in the kitchen dumbwaiter, a vertical transport device.

Lila Mae discerns a relation between the daily anxiety Fulton must have felt and the catastrophic accident of elevator No. 11. "What passing for white does not account for: the person who knows your secret skin, the one you encounter at that unexpected time on that quite ordinary street. What Intuitionism does not account for: the catastrophic accident the elevator encounters on that unexpected moment on that quite ordinary ascent, the one who will reveal the device for what it truly is. The colored man passing for white and the innocent elevator must rely on luck" (231), on the hope that the sister will not appear, the ascent will not be betrayed to gravity.

Lila Mae discovers the elevator's secret when she returns late at night to the Fanny Briggs building. Although No. 11 is now smashed to bits, she re-creates her experience of it with No. 14, using the elevator's ascent to remember, as vividly and in as much detail as possible, her intuitive grasp of No. 11's performance. As she rises, "The genies appear on cue, dragging themselves from the wings. All of them energetic and fastidious, describing seamless verticality to Lila Mae in her mind's own tongue. . . . The genies bow and do not linger for her lonely applause. She opens her eyes. The doors open to the dead air of the forty-second floor. She hits the Lobby button. Nothing" (227).

What Lila Mae confirms in that "Nothing" is that No. 11's demise was not sabotage—not by United or Arbo, Chancre or his underlings. Forensics, she intuits, will find nothing to account for the accident. She likens the catastrophic accident to comets that, having undergone "countless unavailing ellipses," suddenly diverge and hit a planet, "emissaries from the unknowable." She realizes what "her discipline [of Intuitionism] and Empiricism have in common: they cannot account for the catastrophic accident" (228). "The elevator pretended to be what it was not . . . Did it know? After all of Fulton's anthropomorphism: did the machine know itself. Possessed the usual spectrum of elevator emotion, yes, but did it have articulate self-awareness . . . Did it decide to pass? To lie and betray itself? Even Fulton stayed away from the catastrophic accident: even in explicating the unbelievable he never dared broach the unknowable, Lila Mae thinks out of fear" (229). The obscure meaning here is heightened by the free indirect discourse of her conclusion: "This was a catastrophic accident, and a message to her. It was her accident" (229).

CATASTROPHIC FAILURE: THE MEANING OF THE MESSAGE

What message is she discerning? Despite the intelligent criticism this book has attracted from such outstanding critics as Lauren Berlant,

John Johnston, and Ramón Saldívar, among others, to my knowledge no one has published a convincing explanation for this extraordinary passage, or indeed any explanation at all, notwithstanding that it arguably constitutes the narrative's climax. In view of this lacuna, I propose a somewhat controversial interpretation that relies on making a connection between the novel and a classic problem in computational theory. Since the text never mentions computation, this move might ordinarily be considered as overreaching, but I think a strong case can be made for its ability to illuminate a crucial absence in the text. It also makes a provocative connection between a technical device often seen as epitomizing technical cognition—the computer—and another device that is rarely considered in cognitive terms—the electromechanical elevator. The interpretation advanced here thus shows how the notion of a cognitive assemblage may be extended to include not only other technical devices but also overtly political concerns such as racism, gender discrimination, urban infrastructural design, and institutional politics.

John Johnston (2008) gives a valuable hint opening the possibility of this interpretation when he notes in passing, without elaborating, that the elevator is a finite-state machine. A finite-state machine is a device, like an elevator or a turnstile, which has a very limited number of determinate states in which it can operate, with clear transitions between states. A turnstile, for example, is normally in the "locked" position, in which case people are prevented from passing through; after insertion of a coin or token, which initiates a transition phase, it moves to the "unlocked" position, enabling people to pass. Like the turnstile, an elevator has a finite number of states represented by the floors at which it stops; pushing a button initiates a transition phase as it moves from state to state, floor to floor.

The significance of the elevator as a finite-state machine lies in its similarity to another finite-state device, the theoretical computer that Alan Turing proposed as a conceptual device to understand the potentials and limitations of computation. Like the elevator moving from floor to floor but going horizontally rather than vertically, Turing's device has a head that moves along a tape divided into a series of discrete blocks or cells.[2] The head writes a one or zero in a cell or erases what is already there, moving back and forth along the tape the specified number of cells and marking the designated symbol according to its program (or set of instructions). Although the tape has unlimited length, the number of cells it employs is always finite, so it qualifies as a special kind of finite- state machine, albeit with enhanced powers because of its computational potential. By convention, this simple device is called the Turing machine. Turing, along with many others since his seminal 1936–37 article, used it to prove important theorems about computation, and it is widely regarded as foundational to modern computational theory.

In his original publication proposing this device (Turing 1936–37), Turing used it to explore the Entscheidungsproblem or the so-called halting problem.[3] The halting problem is important because it falls in the category of mathematical problems that have been proven to be undecidable. The question Turing examined is this: is there a procedure that, for all possible Turing machine algorithms, can determine in advance whether or not a given algorithm will halt (that is, whether the computation will conclude)? Turing proved that no such general-purpose procedure exists by showing that if it did exist, it would engender a contradiction. The proof is technical, but in broad strokes his strategy was as follows. It is known that most real numbers are not computable, that is, no program exists that can generate them, digit by digit. Turing demonstrated that any program predicting whether a computation would halt would also be able to compute real numbers.[4] Hence, since most real numbers are incomputable, the assumption that such a program would exist must be false. As a result, he established a conceptual limit to what computation can accomplish.

His proof for the halting problem has since been shown to be inter-convertible with Kurt Gödel's incompleteness theorem. Just as Turing proved that the question of whether all possible algorithms will halt is undecidable, Gödel proved that any formal system powerful enough to do arithmetic must have at least one statement that cannot be proven to be either true or false and so is undecidable. Gödel showed that it is always possible to fold statements *about* number theory (that is, meta-statements) into a statement *within* number theory, and the reflexivity resulting from this infolding creates an ambiguity and hence an undecidability. A simple example of a similar ambiguity is the statement "This sentence is false." If the sentence is true, then it must be false; but if it is false, then it must be true: in other words, the statement's truth or falsity is undecidable.

We can now return to the enigmatic passages in which Lila Mae discerns a message in elevator No. 11's failure. Recall that she concludes the catastrophic accident is what cannot be predicted, either by Empiricist or Intuitionist methodology; it is what escapes both the rationality of measurement and the nonconscious cognition of intuition. I propose that the catastrophic failure is the translation into "elevatorese" of the halting problem, rendering the notion of "halting" as a literal cessation of movement (rather than the conclusion of a calculation). By analogy, the question confronting Lila Mae when she revisits the Fanny Briggs building is this: is there a procedure that can decide, in advance, whether a given elevator will halt when it is supposed to? The answer, she realizes, is no: no such procedure exists for all elevators, just as no procedure exists that will determine in advance whether all Turing machine algorithms will halt.

The fact that the probability of such an event is minuscule does not diminish its theoretical importance. Despite the elevator's catastrophic failure being "a million-in-a-million occurrence" (230), its existence is confirmed by the implicit analogy with the halting problem, indicating that it cannot be eliminated completely without engendering a contradiction. Such events, Lila Mae thinks, are "not so much what happens very seldom but what happens when you subtract what happens all the time" (230). "They are, historically, good or bad omens . . . urging in reform . . . or instructing the dull and plodding citizens of modernity that there is a power beyond rationality" (230).

While traditionally the catastrophic failure is the ultimate nightmare for those whose job it is to ensure the safety of vertical transport, so potent a specter that even the visionary James Fulton avoided it, what Lila Mae learns from the catastrophic failure of No. 11 is that another realm beckons beyond the binary choice of Empiricism and Intuitionism: the undecidable.

THE LIBERATORY POTENTIAL OF ERROR

To amplify further the theoretical significance of Turing's work on the halting problem and its relevance to Whitehead's narrative, we can turn to the work of Gregory J. Chaitin (Chaitin 1999; 2001; 2006). Chaitin, fascinated by Turing's proof, asked a related but different question: what is the probability of picking out, at random, a program that will halt from all possible programs the Turing machine can run? Note that Turing's proof did not concern itself with the relative frequency of programs that would halt versus those that would not; it simply asked if it was possible to devise a procedure that could determine, in advance, whether all possible programs would halt. The answer to Chaitin's question was a number that he designated as Omega, and the class of numbers that qualify as Omegas turn out to have unusual properties.

Among them is the fact that the sequence of numbers making up an Omega is random; that is, the prevalence of a given number in the sequence is no different than for that of a fair coin toss. Consequently, since Omegas cannot be computed in totality (that is, the sequence of an Omega cannot be predicted and is infinite), they constitute not just the undecidable but the unknowable. The implications of Omegas for the foundations of mathematics are highly significant, for when tested in terms of number theory, they reveal that even in mathematics, long considered the most exact of the sciences and the foundation for such disciplines as theoretical physics, "randomness is everywhere" (Calude and Chaitin 1999, 319). As these authors conclude, "randomness is as fundamental and as pervasive in pure mathematics as it is in theoretical physics" (320). They continue: "Even after Gödel and Turing showed that Hilbert's dream [that all of mathematics could be shown to be both provable and consistent] didn't work, in practice most mathematicians carried on as before, in Hilbert's spirit. But now, finally, the computer is changing the way we do things. It is easy to run a mathematical experiment on a computer, but you can't always find a proof to explain the results. So in order to cope, mathematicians are sometimes forced to proceed in a more pragmatic manner, like physicists. The Omega results provide a theoretical underpinning for this revolution" (Calude and Chaitin 1999, 401).

Returning to *The Intuitionist,* we find in this revolutionary spirit an explanation adequate to account for Lila Mae's belief that the catastrophic accident and the lesson it teaches her will, along with Fulton's notebooks, open an entirely new terrain that will enable her to make the next great leap forward, the "second elevation" (61). At the beginning of her quest, Mr. Reed had asked her, "What does the perfect elevator look like, the one that will deliver us from the cities we suffer now, these stunted shacks? We don't know because we can't see inside it, it's something we cannot imagine, like the shape of angels' teeth. It's a black box" (61). Reed's imagery suggests that when the black box is opened, the truth will be revealed. The answer Lila Mae intuits, however, is subtler and more powerful. The power of the black box does not lie in concealing a knowable answer, but rather in its symbolization of the limits of knowledge, both Empirical and Intuitionist. The black box cannot be opened because, as an integral and mysterious unity, it gestures toward the unknowable itself.

In volume two of *Theoretical Elevators* (when he began to take seriously the utopianism that began as a joke in volume one), Fulton wrote, "An elevator is a train. The perfect train terminates at Heaven. The perfect elevator waits while its human freight tries to grab through the muck and find the words . . . In the black box, this messy business of human communication is reduced to excreted chemicals, understood by the soul's receptors and translated into true speech" (87). This passage has often been interpreted as an allusion to the kind of intuitive knowledge that Lila Mae had when she did her inspections, but "learning to read" the black box as the unknowable yields a different possibility.

After her enlightenment, Lila Mae remembers the lectures that Fulton gave to "his flock," which they heard but were "not aware of what he [was] truly speaking. *The elevator world will look like Heaven but not the Heaven you have reckoned*" (241). Similarly, what the black box reveals is a truth upon which mathematicians had not reckoned— not the consistent and knowable systems postulated by Hilbert. If for Hilbert Heaven was a completely axiomatized mathematics, this is in contrast a mathematics shot through with randomness, derived as much from intuition as logical deduction and induction, with certain knowledge riddled by unavoidable bursts of the undecidable and unknowable. Meditating on the limits of Fulton's explorations, Lila Mae thinks he "never dared broach the unknowable" because he was afraid (229). When he finally confronted it in the form of the black box, he realized its liberatory potential, but "of course when he started to believe, it was too late" (252); he dies before completing it, leaving to Lila Mae the task of birthing it into the world.

COGNITIVE ASSEMBLAGES AND THE UNKNOWABLE

The Intuitionist is not, of course, a treatise on mathematics but rather a complexly wrought novel in which racism, capitalism, institutions, politics, technical infrastructure, finite-state machines, and the messy human psychologies are entwined, coproducing cognitive assemblages through which choices, interpretations, cognitions, and materialities circulate and coalesce into temporary and shifting configurations and potentialities. To explicate the implicit connections between Turing and Chaitin's work and these assemblages, I turn to an essay by Luciana Parisi, an Italian theorist located at Goldsmiths, University of London, who has brilliantly expounded on the significance of Chaitin's Omegas (Parisi 2015).

Following on Calude and Chaitin's observation that mathematics must now proceed more like the experimental sciences rather than by the supposedly more "rigorous" methodologies of deduction and induction, Parisi turns to the semiotics of C. S. Peirce to explicate the significance of this turn toward what she, following Deleuze and Guattari, calls an "experimental axiomatics" (Parisi 2015, 8). It is typified in her view by Peircean abduction, the process of using data to formulate a best-guess hypothesis. "Chaitin claims that computational processing leads to postulates that cannot be predicted in advance by the program and are therefore to be explained in the terms of 'experimental axiomatics,' a postulate or truth emerging from the inferential synthesis of data carried out by algorithms" (8). This leads in turn to "the emergence of a form of intelligibility able to use data environments— which are the concretizing info-structures of incorporated or automated social practices—to add new axioms, codes, and instructions and new meaning to what was initially programmed. Programming here corresponds to the formation of intelligible procedures in which algorithmic instructions extrapolate new patterns from the data environment they retrieve, thus transforming the pre-established function of programming itself" (8).

In the turn toward algorithms that, instead of excluding contingency and error, learn from them, she sees a major shift in the form of neoliberal capital as it operates through computational media and databases. Rules, rather than governing how the algorithms work, instead emerge abductively from data environments, which themselves may be understood as the "collective use-meaning of data" (3). "This is a mode of reasoning based not on pre-established axioms that need to be proven true," Parisi notes, "but on an hypothetical function, which includes the importance of fallibility or error for the *discovering of new concepts,* involving the revision of both the scientific and manifest image of cognition" (10–11).

In Parisi's explanation, we find an interpretation that will allow us to explain how Lila Mae's "message" might be up to the enormous task not only of reconceiving elevators but of shaking the foundations of entrenched racism and launching the "world's most famous city" on a new trajectory capable of achieving the "second elevation" (Whitehead 1999, 61) of a more just, equitable, and free society. The discovery of the "new concepts" that error and fallibility make possible releases algorithmic reasoning, and by extension the elevator as a finite-state machine, from simply executing the program determining its operation. There is always the possibility of the catastrophic failure, the "thinking through doing" (Parisi 2015, 11) that the elevator enacts through its unexplained, and unexplainable, failure.

This explanation also allows us to understand why nonconscious cognition, symbolized by the sensations received from the elevator and interpreted by core consciousness as visualizations that Lila Mae experiences when inspecting elevators, cannot by itself bring about this transformation. With the advent of affective capitalism and computational media that exploit the missing half-second to hijack human affective responses before consciousness has a chance to evaluate them and respond (Parisi and Goodman, 2011), nonconscious cognition can be held hostage by the designs of neoliberal capital—or in terms of Whitehead's novel, by the institutional racisms and hierarchies that the elevator corporations foster to their advantage. For Parisi, the escape clause that allows a measure of interpretive choice even in the face of affective capitalism goes by the name of general artificial intelligence. "Against the anti-logical machine of neoliberal capitalism ready to deny the autonomy of general intelligence through the affective capture of thinking, a pragmaticist view of reasoning may help us to explain the extent to which fixed capital is not only a new source of surplus value, but also contains an alien logic, extrapolates a new order of meaning that is not readily subsumed under capital's visceral apparatus of capture" (Parisi 2015, 11).

This "alien logic" is very different from that formulated by Mr. Reed in his conversation with Lila Mae, when he implied that the key to transformation may be formulating a theory from the elevator's point of view rather than from the "alien" logic of humans. The elevator he has in mind is the well-performing finite-state machine, not the misbehaving elevator that crashes unexpectedly for no knowable reason. Regardless whether an elevator is seen from an "alien" human view or its own perspective, if it is forever trapped in the discrete finite states of its predetermined journeys, it would be incapable of the kind of radical transformation for which Lila Mae—and we may assume the author—yearns.

The significance of the "experimental axiomatics" that Parisi invokes, following Chaitin, is that they generate new rules and concepts precisely from data environments, which as Parisi notes are the "concretizing info-structures of incorporated or automated social practices" (Parisi 2015, 8). In my terms, data environments are the milieus out of which cognitive assemblages are formed and through which they are able to create new concepts via experimental axiomatics, which in turn change the rules governing how data are processed, which feed back into the cognitive assemblages to transform how they operate. This is the kind of reflexive dynamic that enables cognitive assemblages to evolve in new and unexpected directions—and it is, I suggest, the key to understanding how an elevator's catastrophic failure can expand in widening circles of cognition capable of transforming how the "world's most famous city," and all cities, are constituted and constructed.

In a weird coincidence, Parisi invokes a philosopher—and a philosophy—that has been hovering on the edge of this discussion when she writes that computation may need to be "conceived in terms of its speculative intelligible functions through which unknowns are algorithmically prehended" (Parisi 2015, 8). Prehension is of course the term that Alfred North Whitehead uses to formulate a processual worldview in which "actual entities" (Whitehead 1978, 7, 13 passim) arise and coalesce, a view central to Mark Hansen's reading of twenty-first-century media that operate in temporal regimes inaccessible to humans. The argument advanced here brings together the two Whiteheads, one a novelist who creates a fiction in which an elevator is postulated as an agent capable of prehensions and interpretive choice, and the other the thinker for whom the "extensive continuum" (Whitehead 1978, 61 passim) give rise to prehensions, from which in turn emerge "actual entities." When everything is connected, with everything being influenced by and influencing everything else, transforming elevator technology can potentially transform culture and society.

In this context, errors in an elevator's operations are not mere deficiencies; rather, taken to the extreme of violent and catastrophic failure, they tear open a rip in the temporal fabric of the historical present, through which a better and more utopian future may be glimpsed. As Parisi puts it, what error makes possible is the *"discovering of new concepts*"—in this case concepts about verticality that transform urban architecture, which in turn open new possibilities for rethinking infrastructures, including human, capitalist, technological, and finite-state components as they come together to form new kinds of cognitive assemblages capable of resisting capture by affective and fixed capitalism and transforming the entrenched hierarchies of privilege and the institutionalized racisms associated with them.

AESTHETIC STRATEGIES AND SPECULATIVE REALISM

Even if we disregard (or remain skeptical toward) the novel's allusion to the halting problem for which I have argued, we must nevertheless admit there is something odd about *The Intuitionist*. Located in an unnamed but very recognizable city, set in an unspecified era rendered with considerable historical specificity, *The Intuitionist* is slightly displaced from the history we know—not far enough away to qualify as an alternative history, not close enough to be immediately classifiable as a historical novel. Ramón Saldívar has argued that *The Intuitionist* is part of a literary trend that he characterizes as "postracial" (Saldívar 2013, 1), pointing to Colson Whitehead's op-ed piece in the *New York Times* (Whitehead 2009) in which the writer used the term to characterize American society after Obama's election. Saldívar makes clear from the outset that "race and racism are nowhere near extinct in contemporary America," noting that he follows Whitehead's lead in invoking it "under erasure and with full ironic force" (2).

Nevertheless, he identifies four general characteristics that "postracial" novels exhibit, along with a host of writers and texts representative of the trend. While engaging with postmodern aesthetics, they mix generic forms, focus thematically on race, and qualify as "speculative realism," "a hybrid crossing of the fictional modes of speculative genres, naturalism, social realism, surrealism, magical realism, 'dirty' realism, and metaphysical realism" (5). From my perspective, the most intriguing part of his analysis focuses on why speculative realism should emerge as the aesthetic mode associated with postracial texts. "How can one write the history of the future?" he asks. "What are the conditions of a style appropriate to representing futures that do not and may never exist?" (7). Referring specifically to *The Intuitionist*, he sees it as "the racialized depiction of Utopian desire raised to a second degree, represented by the invention of a new form, a *black noir*, open to the vagaries not of history but of fantasy . . . the point is what this mixing of genres allows for the justification of Utopia, in the face of all evidence to the contrary" (11).

The sense in which he appropriates speculative realism as a literary term shares with the philosophical movement known by that name a desire to break the bounds of finitude—that is break from the closed circle of Kantian thought, in which we are always and inherently distanced from things in themselves, into what Quentin Meillassoux calls the "great outdoors" of other possibilities (Meillassoux 2010, 29). That mathematics, specifically Zermelo-Fraenkel set theory, should be the path into these possibilities had already been explored by Meillassoux, following his teacher Alain Badiou (Meillassoux 2010, 112–28). In this context, it is perhaps not so strange that Whitehead's novel may be linked to the computational theory, specifically the halting problem, as a path into the "great outdoors" of the "second elevation."

Saldívar's argument captures precisely the utopian yearning evident at the novel's conclusion. "They are not ready now but they will be," Lila Mae thinks. "Sometimes in her new room she wonders who will decode the new elevator first. It could be Arbo. It could be United. It doesn't matter. Like the election, their petty squabbling feeds the new thing that is coming. In its own way, it prepares them" (Whitehead 1999, 253). With breathtaking casualness, she dismisses the capitalist enterprise as all but irrelevant, in sharp contrast to the picture that Ben Ulrich, *Lift* investigative journalist and sometime torture victim, had painted for her earlier. "Did you think this was all about philosophy? Who's the better man—Intuitionism or Empiricism? No one really gives a crap about that. Arbo and United are the guys who make the things. That's what really matters" (208).

Ulrich's view fits well with the position argued by Mark Fisher in *Capitalist Realism*, when he reads the film *Children of Men,* with its vision of near-universal sterility, as a metaphor for a society facing the "end of history" (Fisher 2009, 80) that finds impossible the task of imagining an alternative to capitalism. "How long can a culture persist without the new?" Fisher imagines the film asking. "What happens if the young are no longer capable of producing surprise?" (3). He suggests that capitalism, with its remarkable capacity to morph and absorb everything into its dynamics, even putative oppositions and resistances, is akin to *The Thing,* the amorphous, all-devouring entity in John Carpenter's film. Quoting Badiou, he comments, "The 'realism' here is analogous to the deflationary perspective of a depressive who believes that any positive state, any hope, is a dangerous illusion" (5).

Against this backdrop, the peculiarities of *The Intuitionist* can be understood not simply as idiosyncrasies but as politico-aesthetic strategies. Its almost, but not quite our history serves to defamiliarize the history of capitalism just enough to allow a glimmer of hope to enter, while remaining close enough to our historical present to enable us to recognize the structures of inequality and institutionalized racism it depicts. Hopeful but not naïve, not yet fully formed but not vague either, the new era that Lila Mae believes she can midwife into existence trembles at the edge of an unbelievable, yet crucially necessary, affirmation. "It will come. She is never wrong. It's her intuition" (Whitehead 1999, 255).

THE HISTORICAL PRESENT AND COGNITIVE ASSEMBLAGES

Assuming that a path into a better future has been opened, how does it affect the present in which Lila Mae lives, or the present of readers for whom the novel's not quite our present is already not quite our past? What roles do nonconscious cognitions and cognitive assemblages play in the future/past/present transformation? Lauren Berlant provides a helpful framework for addressing these questions in her essay on history, affect, and their interactions (Berlant 2008). She asks, "How is it possible for the affects to sense that people have lived a moment collectively and translocally in a way that is not just a record of ideology?" (846). She posits that affect is the result of "the body's active presence in the intensities of the present," that it "embeds the subject in a historical field," and moreover that "its scholarly pursuit can communicate the conditions of an historical moment's production as a visceral moment" (846). In short, she searches for a way to incorporate affects as historical phenomena responsive to, and partially responsible for, the historical specificities that in retrospect can be recognized as what Raymond Williams called the "structures of feeling" characteristic of a particular era (Williams 1977). As she rightly argues, the Marxist view for which Williams stands as representative has acknowledged that affects must be involved in the production of historical moments, but it primarily emphasizes systematicity and ideology, with the theoretical treatment of affects remaining vague and therefore unsatisfactory.

To create a theory capable of incorporating affect into historical events in more than a superficial way, Berlant employs a term that has already appeared in this chapter, namely the "historical present." She explains, "My interest is in constructing a mode of analysis of the historical present that moves us away from a dialectic of structure (explanation of what is systemic in the reproduction of the world) and agency (what people do in everyday life), and toward attending to their embeddedness in scenes that make demands on the sensorium for adjudication, adaptation, improvisation, and new visceral imaginaries for what the present could be" (846–47). The historical present, then, is "not a matter for retroactive substantialization," but rather "a thing being made, lived through, and apprehended" (848). Moreover, she theorizes that a chronic crisis, and associated literary genres, "produce the present as a constant pressure on consciousness that forces consciousness to apprehend its moment as emergently historic." "Crisis reveals and creates habits and genres of inhabiting the ordinary while reconstituting worlds that are never futures but presents thickly inhabited, opened up, and moved around in" (848). The historical present, then, is at once ordinary and multiple, affectively experienced and yet open to other modes of being that may also be inhabited.

Berlant's essay is a wonderful analysis, so good it makes your teeth ache to read it. When she turns to *The Intuitionist,* one of her two tutor texts (the other being William Gibson's *Pattern Recognition*), she astutely recognizes the important role that affect plays in the text, quoting at length Lila Mae's intuitive evaluation of the elevator at 125 Walker. The weak point comes, however, when Berlant arrives at the catastrophic failure of elevator No. 11. She reads Lila Mae's late-night visit to the Fanny Briggs building as "the elevator calling out to her to clear her name and find a higher truth" (853). Although she recognizes that this somehow involves a retraining of Lila Mae's intuition, she sees this as an "intuitional shift to living a fearless racialized imaginary in the present, with no fidelity toward protecting the built white world as such, and theorizing the beyond as an act of vitalism" (854), with no explanation for how a catastrophic accident is able to launch Lila Mae on this path or what kind of "higher truth" might be involved. Crucially for her argument, she has no way to connect the *personal* message that the elevator's failure reveals to Lila Mae with the larger *social* transformation of the impending change that Lila Mae's efforts will bring about—no way, that is, to move from affect felt and experienced by a single person to larger social experiences of affect that construct and define the specificities of the historical present.

My previous argument about the halting problem provides precisely the kind of connections and links that Berlant's essay is missing. It relates the personal message Lila Mae receives to broader social concerns and indeed to the nature of epistemology itself, especially the productive roles of error and failure in an era on the verge of developing the powerful computational media that would profoundly alter the dynamics of human-technical cognitions. I want to underscore the importance of affects in cognitive assemblages, a matter on which Berlant's essay is admirably eloquent. She shows with forceful clarity how affect can thicken and extend into prehension of historical events, without however becoming an "eventilization" (849), sliding into the past, and thereby losing its potency. As she argues, they continue to operate in the "historical present," rendering it at once as immediate lived experience and as prehensions that interact with other contingencies to form a historical nexus.

Recall that nonconscious cognition is intimately related to affect, serving as a site of mediation between bodily and visceral actions and the higher modes of consciousness/unconsciousness. In this context, the theoretical importance of cognitive assemblages stems from their ability to show how affects come together with other forms of technical and human cognitions to create dynamic systems flexible enough to change their configurations continuously, and stable enough to function within the complex architectures of human-technical interactions. Within such systems, errors and contingencies may, at decisive cusp points, tip the system one way or another so that the assemblage begins to operate in new and unanticipated modes.

COGNITIVE ASSEMBLAGES AND NOVELISTIC FORMS

Berlant's essay strives to incorporate affects into history and therefore into our understanding of historical (or, in the case of *The Intuitionist,* almost historical) novels. Behind her quest looms another question relevant to my project here: what does literature, especially the novel, contribute to our understanding of cognitive assemblages and the roles played by human and technical nonconscious cognitions? What specific dynamics do novels enact that are not already present in sociology, history, philosophy, human-computer interface design, statistics, mathematics, physics, and biology, for example? For convenience I use examples from *The Intuitionist,* but most novels would offer similar instances. A tentative answer would include the following points:

1. *Novels show how affects work in individual and collective human lives.* The precariousness of Lila Mae's position in the "world's most famous city" is shown not only through the suspicious reactions and petty meanness she experiences but through how it affects her body, dress, emotions, gestures, postures, facial expressions. She lies in bed at night moving the muscles of her face to arrive at exactly the right expression to show to a hostile and indifferent world, a fragile defense against slurs, insinuations, unwanted advances, disrespect. On the collective side, the scene of the elevator inspectors gathered at O'Connor's bar (Whitehead 1999, 24) to watch the news about the catastrophic failure of No. 11 shows how the excitement of the crowd is contagious, spurring the watchers on through a sense of group solidarity and its necessary other, the exclusion of those who, like Lila Mae, do not belong. When Lila Mae corrals Chuck, her one and only friend in the department, into the ladies room to enlist his help (34–37), his discomfort at the setting is exacerbated by the drunk woman passed out on the room's only toilet in a posture that reminds him of his mother. These affects go a long way to explaining why Chuck agrees to help Lila Mae, despite the risk he runs.

2. *Novels provide specific contexts—historical, racial, gendered, economic, psychological—of lived experiences.* The basement of Johnny Shush's unsafe (rather than safe) house doubles as a torture chamber, and the mattress and wall stains that Ben Ulrich sees there speak volumes about the room's purpose and former occupants (94–95). When Lila Mae is also taken to the basement, the implicit menace increases the tension of her conversation with Chancre. It also makes believable his assertion that he did not sabotage elevator No. 11, for clearly he has the upper hand and therefore no reason to lie to Lila Mae. Ben Ulrich later tells Lila Mae that he knew his abductors were not Johnny Shush's boys, although that is what they pretended, but rather were corporate stooges, from the quality of the shirts they wore (210). Details of architecture, clothing, furniture, and myriad other small specificities flesh out and give meaning to the reactions of characters and consequently of readers as well.

3. *Novels depict ranges of interpretations and choices that drive the dynamics of systems.* When Lila Mae visits Fulton's housekeeper, Marie Claire Rogers, a second time, her house (and it is hers, she had reminded Lila Mae on her first visit) has been desecrated by corporate goons, who not only broke all her ceramic horses decorating the mantle but also urinated on the floor. When Lila Mae picks up a broom to help and lies about the room not smelling of urine, these small courtesies may be what tip the balance and make Mrs. Rogers decide to give her Fulton's remaining papers. Lila Mae's choice to help rather than stand by creates a small perturbation in the system that nevertheless initiates a large change in its dynamics.

4. *Novels provide form and shape experience.* Berlant's essay defines genre as a "loose affectual contract that predicts the form that an aesthetic transaction will take" (Berlant 2008, 847). The hybridity of Whitehead's novel stems, as Saldívar points out, from its mixing of realist modes of narration, setting, character, and action with the fantastical elements of intuiting an elevator's health by making contact between a human back and the elevator's vibrations, of attributing to the elevator emotion, the capacity to lie, and the ability to engage in deliberate misdirection as a form of elevator "passing." The kinds of aesthetic transactions the novel's form predicts, guides, and enacts, then, take place in a borderland between the hard realities of racial and gendered structural inequalities and the speculative possibilities of utopian transformation. Contingency, error, and necessity mingle indiscriminately, with each playing a role in determining how the dynamics of cognitive assemblages emerge and evolve. Fulton sees Lila Mae on campus and asks her name; then, distracted by the cost of resoling his shoes and other matters, he scribbles in the margin of his paper, "Lila Mae is the one" (251), a message hovering indeterminately between prophecy and coincidence, error and anointment.

5. *Novels enact connections and links between disparate phenomena.* The entwining of racial uplift and vertical transport, notable in Fulton's papers but present elsewhere as well, provides the ground for a whole series of other analogies—a black man passing for white and a faulty elevator passing for healthy; a second elevation in architecture and race relations; a technological revolution that will rearrange cities and souls. The enactment of unlikely connections is typified in the scene where Lila Mae, studying in her room at the

Institute late at night, has a clear line of sight to James Fulton as, already slipping into dementia, he moves about the library located at the Institute's top floor late at night. He sees her light, she sees his in a contingent arrangement that, in retrospect, is an uncanny coincidence or the beginning of a mystical bond between them.

6. *Novels use literary resources to mean more than they say.* On a thematic level, the allusions to the halting problem and the black box as the unknowable show how the range of reference can expand in almost unlimited fashion. In literary terms, rhetorical tropes such as irony, metaphor, metonymy, and synecdoche (Kenneth Burke's four master tropes) show how literary language can function affectively as well as conceptually. Garrett Stewart (1990) has argued that literary language differs from ordinary prose precisely in its ability to address a wide range of embodied responses. If we ask not what but where we read, Stewart comments, the answers would include: in the throat through subvocalization, the viscera through embodied responses, the circulatory system through increases or decreases in blood pressure, the central nervous system through pupil contraction and dilation in response to suspenseful or peaceful passages, and a host of other embodied and affective reactions.

7. *Novels explore ethical issues in specific and concrete, rather than abstract, terms.* Lila Mae's decision to give Fulton's notebook to Raymond Coombs is fraught with irony, from her opening confrontation in his Arbo office to her parting shot, "I just wanted to help" (251). What he does not yet know is that she also gave it to Chancre and thus to Arbo's archcompetitor United, and to Ben Ulrich, the investigative reporter who has positioned himself as the enemy of both corporations. Her equal-opportunity disclosures indicate how irrelevant the recipients' differences have become in her mind as she contemplates the coming future. "The elevator she delivered . . . should hold them for a while. Then one day they will realize it is not perfect. If it is the right time she will give them the perfect elevator. If it is not time she will send out more of Fulton's words to let them know it is coming" (255). Her choices, at once ethical, political, and technological, indicate the shift of mindset that has positioned her as the leader and designer of the future, rather than as someone who can at best only react to actions that others take. Her allegiance—we might say, her only allegiance—is to the future she is birthing and the utopian possibilities it harbors.

The above points refer, of course, to representations *within* novels, but novels also function as cognitive devices in larger assemblages that include publishers, readers, reviewers, media, networked and conventional dissemination channels, and a host of other human and technical systems loosely aggregating to form flexible and shifting cognitive assemblages through which choices, interpretations, and contexts operate as information flows through and between systems. The broader significance of intuition comes not only from its contrast with empirical rationality (as the mutually entailed limitations of Intuitionism and Empiricism in Whitehead's novel suggest), but from the assemblages in which cognizers at many levels cooperate and compete to create the emergent dynamics of human-technical interactions. Nonconscious cognition is an important key to understanding how these assemblages form and transform, especially in the connections it builds between human and technical cognizers, but it is in and through cognitive assemblages themselves, in the folds in time (Serres and Latour 1995; Latour 1992, 230) they create in which the past jostles with the future, that the historical present comes thickly into existence.

CHAPTER 8

The Utopian Potential of Cognitive Assemblages

At midcentury, Norbert Wiener was struggling with what he saw as the peril and promise of the cybernetic paradigm. One result of that struggle was *The Human Use of Human Beings* (1950), which is less a coherent argument than a somewhat chaotic mix of hope and dread. Half a century later, we can see with the benefit of hindsight in what ways the cybernetic paradigm was both prophetic and misguided. It was correct in anticipating that modes of communication between humans, nonhuman life-forms, and machines would come to be increasingly critical to the future of the planet; it was wrong in thinking that feedback mechanisms were the key to controlling this future. In fact the whole idea of control, with its historical baggage of human domination and exceptionalism, has come to seem increasingly obsolete, if not outright dangerous. Now well into the new millennium, we can appreciate the enormous differences that networked and programmable media have made in human complex systems, and we are beginning to glimpse how these conditions have opened new possibilities for utopian thoughts and actions.

If control in the sense of anticipating all relevant consequences and using this foreknowledge to determine the future has been consigned to the dustbin of history, its demise reveals that the very attempts to render formal (mathematical and computational) systems tractable by rigorous procedures defining boundaries and establishing protocols have confirmed the existence of what lies beyond those boundaries: the incomputable, the undecidable, and the unknowable. Luciana Parisi, in her work on general artificial intelligence (2015), points to the importance of Gregory Chaitin's work in this regard, as discussed in chapter 7, and she expands on the liberatory potential it offers for developed societies, where correlations between databases and increasingly sophisticated surveillance techniques seem to make the interacting imperatives of state control and capitalist exploitation ever more intrusive and oppressive. Beatrice Fazi (2015), who completed her dissertation under Parisi's direction, approaches the issue from another direction, showing how Turing's work on incomputable numbers opens an area of incomputability within the regime of computation itself. In brief, what this work reveals is that the more control is codified and extended through computational media, the more apparent it becomes that control can never be complete, and the very operations that make control possible also authorize its antithesis, areas where the unknowable rules.

The problem, then, is how to use this potential to make real differences in the real world. In my view, this is what motivates Mark B. N. Hansen's work (2015) on Whitehead's philosophy in relation to twenty-first-century media. He adopts from Whitehead the idea of a world continuously in flux; he modifies Whitehead precisely to forge a connection between this flux and "superjects," the settled entities that congeal out of the flux. His point is that processual dynamics does not cease where the human begins, but continues to interpenetrate it, opening new possibilities for resistance and intervention.

My own contribution has focused on the importance of cognition, interpretation, and choice, and the resulting formation of cognitive assemblages in which human and technical actors communicate and interact on many levels and at multiple sites. The complexity of these assemblages, for example in finance capital, has clearly shown that control in the sense Wiener evokes it is no longer possible. Cognition is too distributed, agency is exercised through too many actors, and the interactions are too recursive and complex for any simple notions of control to obtain. Instead of control, *effective modes of intervention seek for inflection points at which systemic dynamics can be decisively transformed to send the cognitive assemblage in a different direction.*[1] For Brad Katsuyama, the inflection point consisted of altering the speed at which algorithmic transactions could be conducted. For the authors of the batch auction proposal, it was devising techniques that made capitalist competition focus on price rather than speed. For ATSAC, it was making the urban traffic infrastructure a site of cooperation between human interventions and intelligent algorithms. For Rosi Braidotti in her quest for an affirmative posthumanism, it was finding the bal-

ance point between flux and stability, human identity and the forces
that interpenetrate and destabilize it. For Colson Whitehead, it was
using a catastrophic elevator failure to show that the future can never
be entirely determined by a past burdened with institutional racism,
hatred, and suspicion.

Thinking about what the very different agendas of Parisi, Fazi, Hansen, Katsuyama, Braidotti, and C. Whitehead (and many others, too
numerous to mention here) have in common leads to useful generalizations about the kind and scope of interventions that can make
differences in real-world systems. First, all these thinkers, activists,
and writers spent the time and conceptual resources necessary to understand the system in detail, whether it is computational regimes,
HFT, processual philosophy, institutional racism, or posthumanist
studies. Only if the system in question is interrogated closely and re-
searched thoroughly can the inflection points be located. Second, once
the inflection points are determined, the next issue is how to introduce
change so as to transform the systemic dynamics. Third, and perhaps
most important, these theorists, activists, and writers draw upon prior
visions of fair play, justice, sustainability, and environmental ethics to
determine the kinds of trajectories they want the system to enact as a
result of their interventions. These are typically not found within the
system itself but come from prior commitments to ethical responsibilities and positive futures.

This is why I have been urging throughout that the humanities have
vital roles to play in thinking about cognitive assemblages. Interpretation, meaning, and value, while not exclusively the province of the humanities, have always been potent sites for exploration within the humanities, including art, literature, philosophy, religious studies, and
qualitative history. Ethics cannot be plastered on as an afterthought
after the system has already been formed and set in motion, an un-
fortunate tendency, for example, in courses on "ethics" in business
practices, which too often focus on how to satisfy legal requirements
so that one does not become the object of a lawsuit. On the contrary,
effective ethical intervention has to be intrinsic to the operation of the
system itself.

For cognitive assemblages, this means becoming knowledgeable
about how the interpenetrations of human and technical cognitions
work at specific sites, and how such analyses can be used to identify
inflection points, which, rather than preexisting as objective realities,

emerge in interaction/intra-action with prior commitments to create new trajectories for the assemblages, providing more open, just, and sustainable futures for humans, nonhuman life-forms, and technical cognizers—which is to say, for the planetary cognitive ecology.

For these utopian possibilities to be realized, humanities scholars must recognize that they too are stakeholders in the evolution of cognitive assemblages, which implies an openness toward learning more about the computational media at the heart of cognitive technical systems. At present, the digital humanities are the contested sites where these issues are discussed and debated, sometimes in heated and angry exchanges. The following section expresses my take on these debates and argues for a more constructive dialogue between the traditional and digital humanities.

ENLARGING THE MIND OF THE HUMANITIES

During a stint at the University of Chicago as the *Critical Inquiry* visiting professor, I was invited to give a presentation to the Society of Fellows. I began my presentation by recalling a digital case study I had conducted (with Allen Riddell) on the constraints operating in Mark Danielewski's elaborately patterned novel *Only Revolutions* (Hayles 2012). As I discussed the project, one of the participants objected that the computer algorithms we employed could deal only with "dumb" questions, not interesting interpretive ones. I responded that the "dumb" answers led to interesting interpretive possibilities, for the absence of certain words was a strong indicator of the constraints the author had imposed, which in turn led to questions about what these constraints implied. My interlocutor continued to insist that ambiguities were the essence of literary interpretation, and that the kind of "distant reading" we had done was reductive and therefore not really humanistic at all. Since she was obviously intelligent and passionate about her position, I made a point of talking with her afterward to explore more fully the nature of her objections. She summed them up by remarking, "It all depends on the kind of humanities you want."

Her passion made me realize that many scholars choose to go into the humanities because they do not like the emphasis in the sciences on finding answers to well-defined questions. Indeed, they tend to believe that interesting questions do not have definite answers at all, offering instead endless opportunities for exploring problematics. They fear that if definite answers were established, interpretation would be shut down and further research would be funneled into increasingly narrow avenues. Leaving aside the question of whether this is an accurate or fair view of scientific investigation, I think it captures the spirit of what my interlocutor meant by "the kind of humanities" she wanted. If computer algorithms could establish definite answers (such as whether or not a certain word appeared in a text, and if so how frequently), then for her and like-minded scholars, the open space that the humanities has established for qualitative inquiry as a bulwark against quantitative results was at risk of crumbling, and all that would be left would be studies dominated by quantitative measures.

Entangled with this attitude are many questions about the proper mission for the humanities, especially in relation to the sciences, and the strategies that humanists employ that may be considered distinctive and so indicative of their contributions to contemporary intellectual life. For a very long time, scholars in the humanities have felt threatened and underappreciated relative to more powerful and culturally central fields, and these perennial concerns are now being exacerbated with the emergence of the digital humanities.

In my view, the digital humanities ought not to be seen as a threat but as an important ally to the traditional humanities, expanding their influence and widening the scope of what counts as humanistic inquiry without sacrificing their distinctive contributions. Moreover, I see the recognition of nonconscious cognition in humans and technical systems as key to repositioning the humanities at the very center of contemporary intellectual inquiry. To make this case, I will discuss the interactions between description and interpretation, clarify the role they play in constructing computer algorithms, identify cultural productions where nonconscious cognitions are being staged as artistic projects, and speculate on the future of human and technical cognitions as they interact through cognitive assemblages. Having shown in previous chapters the centrality of nonconscious cognition in human and technical assemblages, I will in this final chapter carry the argument into the heart of my own discipline and intellectual commitments.

INTERPRETATION AND DESCRIPTION

In the humanities, interpretation is typically regarded as having a higher value than description, a schema with implications for how one regards computational methods. While no one doubts that a word frequency algorithm can count words accurately, many believe that this does not count as a cognitive activity. Interpretation, by contrast, is often seen as an exclusively human prerogative and highly valued as a result. What would it imply, then, to claim that the cognitive nonconscious interprets?

The implications for humanistic strategies are profound. Interpretation is deeply linked with questions of meaning; indeed, many dictionaries define interpretation in terms of meaning (and meaning in terms of interpretation). Meaning, in turn, is central to the mission of humanistic disciplines. Whereas scientific fields always ask "What is it?" and frequently query "How does it work?" they seldom ask why things are as they are, and even less often what they mean. By contrast the humanities, including art history, religious studies, philosophy, history, and literary studies, among others, take the quest for meaning to be central. Why study history, for example, if not to try to determine why events proceeded as they did, and what it means that they did so? Of course, the humanities are not so naïve as to suppose that meaning is easily recoverable or even that it exists other than as human fantasy. From *Oedipus Rex* to *Hamlet* to *Waiting for Godot* and beyond, literary art has confronted the possibility that there may be no transcendent meaning for human life. Nevertheless, even to answer the quest in the negative still involves meaning making as a central problematic.

Until the twentieth century, meaning and interpretation focused primarily on consciousness and its meditations. With the advent of Freudian psychoanalysis, the unconscious was explicitly articulated as another player in the creation of meaning and interpretation, and as a result, earlier and contemporary literary works were reread for their psychoanalytical import. The fact that they could be so reinterpreted implies that the unconscious was always intuited as an important component of human thought. In an important sense, Freud did not so much invent the unconscious as discover it, drawing in part on literary representations that powerfully depict it in action. Now the humanities are being confronted with other major players: human and technical nonconscious cognitions. To engage productively with them, the humanities must broaden their concepts of meaning and interpretation to include such functionalities as recognizing patterns, drawing inferences from those patterns, learning nonconsciously, and correlating complex variables as they change in relation with one another.

Not coincidentally, these functionalities are often used to describe the work done by computer algorithms. A common perspective in the humanities is that these activities are far inferior to what human consciousness can do: John Guillory speaks for many when he says that there is at present an "immeasurable" gap between literary interpretation and what computer algorithms can accomplish (Guillory 2008, 7). That perception is why it is critically important for the humanities to become aware of how nonconscious cognition operates both in human brains and computational media.

This alone would be a compelling reason for the humanities to rethink how meaning and interpretation work in nonconscious processes, a reorientation equivalent in scope, magnitude, and implication to the seismic shock initiated by the explicit recognition of the unconscious. In addition, the humanities are being impacted as never before by computational media, especially in the digital humanities. Once cognition is recognized as operating nonconsciously as well as consciously, a vast array of social, cultural, and technological issues come into view as appropriate for humanistic inquiry. As argued in earlier chapters, these range from interactions of human cognition with the nonconscious cognitions of technical systems to the social, cultural, and economic formations that enter into these interactions through cognitive assemblages.

Of course, it is still possible that some in the humanities may choose to ignore these questions and the possibilities they open. Some may remain content with traditional views that locate meaning and interpretation solely within the human consciousness and unconscious. These views are not so much false as incomplete. To compensate for this limited scope, such scholars should in my view recognize that nonconscious cognition operates within human neurology and appreciate the capabilities it possesses. One possibility that this recognition opens is that the cognitive nonconscious may be discovered in older texts as well as recent works. Chapters 4 and 7 model a few reading strategies for interpreting representations of nonconscious cognition and for exploring the costs of consciousness, but many more possibilities exist, from measuring indicators of nonconscious processing in readers (Riese et al. 2014) to investigating the interplay between affectual responses and larger cognitive assemblages.

For those opposed to the digital humanities, the charge is often made that all algorithms can do is describe, a characterization that, in the prevailing value schema, automatically relegates them to a lower strata and therefore not the "real" or "important" humanities. The clear binary thus established between description and interpretation is open to objections on multiple counts. Science studies, for example, has long recognized that description is always theory laden, because every description assumes an interpretive framework determining what details are noticed, how they are arranged and narrated, and what interpretations account for them. Sharon Marcus, answering critics who contest her and coauthor Stephen Best's call for "surface reading," confronts the entanglement of interpretation with description head-on (Marcus 2013). Rather than arguing that description is not theory laden, Marcus turned the tables by pointing out that every interpretation necessitates description, at least to the extent that descriptive details support, extend, and help to position the interpretation. Although not the conclusion she draws, her argument implies that description and interpretation are recursively embedded in one another, description leading to interpretation, interpretation highlighting certain details over others. Rather than being rivals of one another, then, in this view interpretation and description are mutually supportive and entwined processes.

This helps to clarify the relation of the digital humanities to traditional modes of understanding such as close reading and symptomatic interpretation. Many print-based scholars see algorithmic analyses as rivals to how literary analysis has traditionally been performed, arguing that digital humanities algorithms are nothing more than glorified calculating machines. But this implication misunderstands how algorithms function. Broadly speaking, an algorithmic analysis can be either confirmatory or exploratory. Confirmatory projects are often misunderstood as simply repeating what is already known, for example, what literary dramas fall into what genre category (see Moretti 2013 for a superb example of this kind of analysis). The point, however, is not to determine, for example, what literary drama falls into what generic category, but rather to make explicit the factors characterizing one kind of dramatic structure rather than another. Often new kinds of correlations appear that raise questions about traditional criteria for genres, stimulating the search for explanations about why these correlations pertain. When an algorithmic analysis is exploratory, it seeks to identify patterns not previously detected by human reading, either because corpora are too vast to be read in entirety, or because long-held presuppositions constrain too narrowly the range of possibilities considered.

Just as interpretation and description are entwined for human readers (as Marcus's argument implies), so interpretation enters into algorithmic analyses at several points. First, one must make some initial assumptions in order to program the algorithms appropriately. In the case of Tom Mitchell's Never-Ending Language Learning (NELL) project at Carnegie Mellon, the research team first constructs ontologies to categorize words into grammatical categories (Mitchell n.d.). In Timothy Lenoir and Eric Gianella's algorithms designed to detect the emergence of new technology platforms by analyzing patent applications (Lenoir and Gianella 2011), they reject ontologies in favor of determining which patent applications cite the same references. The assumption here is that cocitations will form a network of similar endeavors, and will lead to the identification of emerging platforms. Whatever the project, the algorithms reflect initial interpretive assumptions about what kinds of data are likely to reveal interesting patterns. Stanley Fish to the contrary (Fish 2012), there are no "all-purpose" algorithms that will work in every case.

Second, interpretation strongly comes into play when data are collected from the algorithmic analysis. When Matthew Jockers found that Gothic literary texts have an unusually high percentage of definite articles in their titles (discussed in McLemee 2013), for example, his interpretation suggested this was so because of the prevalence of place names in the titles (*The Castle of Otranto,* for example). Such conclusions often lead to the choice of algorithms for the next stage, which are interpreted in turn, and so forth in recursive cycles.

Employing algorithmic analyses thus follows a similar pattern to human description/interpretation, with the advantage that nonconscious cognition operates without the biases inherent in consciousness, where presuppositions can cause some evidence to be ignored or underemphasized in favor of other evidence more in accord with the researcher's own presuppositions. To take advantage of this difference, part of the art of constructing algorithmic analyses is to keep the number of starting assumptions small, or at least to keep them as independent as possible of the kinds of results that might emerge. The important distinction with digital humanities projects, then, is not so much between description versus interpretation but rather the capabilities and costs of human reading versus the advantages and limitations of technical cognition. Working together in recursive cycles, human conscious analysis, human nonconscious cognition, and technical cognition can expand the range and significance of insights beyond what each can accomplish alone.

STAGING THE COGNITIVE NONCONSCIOUS
IN THE THEATER OF CONSCIOUSNESS

If my hypothesis is correct about the growing importance of nonconscious cognitions and the cognitive assemblages in which they operate, we should be able to detect these influences in contemporary literature and other creative works. Of course, since these products emerge from conscious/unconscious modes of awareness, what will be reflected is not the cognitive nonconscious in itself, but rather its restaging within the theater of consciousness. One of the sites where this staging is readily apparent is in contemporary conceptual poetics. Consider, for example, Kenneth Goldsmith's "uncreative writing." In *Day*, Goldsmith retyped an entire day (September 1, 2000) of the *New York Times*; in *Fidget*, he recorded every bodily movement for a day; in *Soliloquy*, every word he spoke for a week (but not those spoken to him); and in *Traffic*, traffic reports, recorded every ten minutes over an unnamed holiday, from a New York radio station. His work and accompanying manifestos have initiated a vigorous debate about the work's value. Who, for example, would want to read *Day*? Apparently not even Goldsmith, who professed to type it mechanically, scarcely even looking at the page he was copying. He often speaks of himself as mechanistic (Goldsmith 2008), and as the "most boring writer who ever lived" (qtd. in Perloff 2012, 149). In his list of favored methodologies, the parallel with database technologies is unmistakable, as he mentions "information Management, word processing, databasing, and extreme process . . . Obsessive archiving & cataloging, the debased language of media & advertising; language more concerned with quantity than quality" (Goldsmith 2008). Of course we might, as Marjorie Perloff does, insist there is more at work here than mere copying (Perloff 2012). Still, the author's own design seems to commit him to enacting something as close to Stanley Fish's idea of algorithmic processing as humanly possible—rote calculation, mindless copying, mechanical repetition.

It seems, in other words, that Goldsmith is determined to stage nonconscious cognition as taking over and usurping consciousness, perhaps simultaneously with a sly intrusion of conscious design that a reader can notice only with some effort. That he calls the result "poetry" is all the more provocative, as if the genre most associated with crafted language and the pure overflow of emotion has suddenly turned the neural hierarchy upside down. The irony, of course, is that the cognitive nonconscious is itself becoming more diverse, sophisticated, and cognitively capable as it extends into technical systems. Ultimately what is mimed here is not the actual cognitive nonconscious but a parody version that pulls two double-crosses at once, at both ends of the neuronal spectrum: consciousness performed as if it was nonconscious, and the nonconscious performed according to criteria selected by consciousness. As Perloff notes, quoting John Cage, "If something is boring after two minutes, try it for four. If still boring, try it for eight, sixteen, thirty-two, and so on. Eventually one discovers that it's not boring at all but very interesting'" (157). Consciousness wearing a (distorted) mask of the cognitive nonconscious while slyly peeping through to watch the reaction—that's interesting!

Another example of how the cognitive nonconscious is surfacing in contemporary creative works is Kate Marshall's project on contemporary novels, which she calls "Novels by Aliens." Focusing on "the nonhuman as a figure, technique and desire," Marshall shows that narrative viewpoints in a range of contemporary novels exhibit what Fredric Jameson calls the "ever-newer realisms [that] constantly have to be invented to trace new social dynamics" (Jameson 2010, 362). In Colson Whitehead's *Zone One,* for example, the viewpoint for the Quiet Storm's highway-clearing project involves an overhead, faraway perspective more proper to a high-flying drone than to any human observer (Marshall 2014). The protagonist, nicknamed Mark Spitz, collaborates with Quiet Storm in part because he feels, as Marshall puts it, "lust to be a viewpoint." Although Marshall links these literary effects to such philosophical movements as speculative realism, it is likely that both speculative realism and literary experiments in nonhuman viewpoints are catalyzed by the expansive pervasiveness of the cognitive nonconscious in the built environments of developed countries. In this view, part of the contemporary turn toward the nonhuman is the realization that an object need not be alive or conscious in order to function as a cognitive agent.

TWO PATHS FOR THE HUMANITIES

Today the humanities stand at a crossroads. On one side the path continues with traditional understandings of interpretation, closely linked with assumptions about humans and their relations to the world as represented in cultural artifacts. Indeed, the majority of interpretive activities within the humanities arguably have to do specifically with the relation of human *selves* to the world. This construction assumes that humans have selves, that selves are necessary for thinking, and that selves originate in consciousness/unconsciousness. The other path diverges from these assumptions by enlarging the idea of cognition to include nonconscious activities. In this line of reasoning, the cognitive nonconscious also carries on complex acts of interpretation, which syncopate with conscious interpretations in a rich spectrum of possibilities.

What advantages and limitations do these two paths offer? The traditional path carries the assumption that interpretation, requiring as it does consciousness and a self, is confined largely if not exclusively to humans (perhaps occasionally extended to some animals). This path reinforces the idea that humans are special, that they are the source of almost all cognition on the planet, and that human viewpoints therefore count the most in determining what the world means. The other path recognizes that cognition is much broader than human thinking and that other life-forms as well as technical devices cognize and interpret all the time. Moreover, it also implies that these interpretations interact with and significantly influence the conscious/unconscious interpretations of humans, which themselves depend on prior integrations and interpretations by human nonconscious cognitions. The search for meaning then becomes a pervasive activity among humans, animals, and technical devices, with many different kinds of agents contributing to a rich ecology of collaborating, reinforcing, contesting, and conflicting interpretations.

One of the costs of the traditional path is the isolation of the humanities from the sciences and engineering. If interpretation is an exclusively human activity and if the humanities are mostly about interpretation, then there are few resources within the humanities to understand the complex embeddedness of humans in cognitive technical environments and in relationships with other species. If, on the contrary, interpretation is understood as pervasive in natural and built environments, the humanities can make important contributions to such fields as architecture, electrical and mechanical engineering, computer science, industrial design, and many others. The sophisticated methods that the humanities have developed for analyzing different kinds of interpretations and their ecological relationships with each other then pay rich dividends and open onto any number of exciting collaborative projects.

Proceeding down the nontraditional path, in my view much the better choice, requires a shift in conceptual frameworks so extensive that it might as well be called an epistemic break. As we have seen, one of the first moves is to break the equivalence between thought and cognition; another crucial move is to reconceptualize interpretation so that it applies to information flows as well as to questions about the relations of human selves to the world. With the resulting shifts of perspective, many of the misunderstandings about the kinds of interventions the digital humanities are now making in the humanities simply fade away. I want to emphasize that the issues involved here are much larger than the digital humanities in themselves. Important as they are, focusing only on them distorts what is at stake (which is one reason why I have waited until this final chapter to introduce the topic). The point, as far as I am concerned, is less about methods that seem to be rivals to interpretation—a formulation that assumes "interpretation" and "meaning" are stable categories that can be adequately discussed as exclusively human activities—than it is about the scope and essence of cognition as it operates in humans and technical systems, and in the larger cognitive assemblages that are transforming the planet's built and natural environments.

There are two important implications that I want to bring out as a conclusion to this chapter (and book). The first is that nonconscious cognition, far from being fundamentally alien to how humans think, is in fact crucial to human cognition, as we have seen in chapter 2, summarizing research investigating the relation between consciousness and nonconscious cognition. Those in the humanities who see an "immeasurable" gap separating what computers can do from what human brains achieve should rephrase their argument to take into account that low-level nonconscious cognitive activities are always already involved in high-level thoughts in human brains. In the digital humanities and many other sites, external nonconscious cognizers are being enrolled in the human extended cognitive system, just as historically humans have excelled as a species in enrolling all manner of external objects as cognitive supports and extensions (Clark 2008; Hutchins 1996). These external cognizers perform tasks that also take place within human brains, including recognizing patterns, drawing inferences from complex arrays of data, learning to recognize covariation among multiple variables, and reaching decisions about conflicting or ambiguous information.

Seen in this way, computers are not creatures alien to humans, as they are sometimes depicted in popular culture and in some popular science books. For example, David Eagleman in *Incognito* (2012), a popularized account of research into nonconscious cognition, constantly refers to specialized automated processors within the human brain as "alien" and "zombie" systems. Such rhetoric, no doubt fashioned to make his argument seem more lively and entertaining, introduces a totally artificial division between consciousness and nonconscious cognition, as if some miniature computer resides in the human brain that is fundamentally removed from and alien to the self. In fact, however, the brain is an amazing integrated system in which every part communicates and is integrated with every other part (as Eagleman himself recognizes in another context [166]), and in which nonconscious cognition needs the support of high-level amplification signals to endure, no less than consciousness depends on and integrates the fast-response information processing of nonconscious cognitions.

This leads to the second point I want to emphasize. We are now in a period of increasing complexity, sociality, and interconnections between technical nonconscious systems. Just as human cognition was given a sharp evolutionary boost upward through human sociality, so nonconscious cognition in technical systems operates not in isolation but in recursive interconnections with other technical systems. In the "Internet of things," for example, a nonconscious system such as VIV, discussed in chapter 5, may have access to open-source information on the web, which it can use to create connections, leveraging its inferences from individual sites through cross-connections that leap to still further inferences, and so on. Biological organisms evolved consciousness to make this kind of quantum leap from individual instances to high-level abstractions; core and higher consciousness in turn ultimately enabled humans to build sophisticated communication networks and informational structures such as the web. In large-scale historical perspective, automated cognizers are one result of evolved human consciousness.

It is likely, however, that the evolutionary development of technical cognizers will take a different path from that of *Homo sapiens.* Their trajectory will not run through consciousness but rather through more intensive and pervasive interconnections with other nonconscious cognizers. In a sense, they do not require consciousness for their operations, because they are already in recursive loops with human consciousness. Just as from our point of view they are part of our extended cognitive systems (Clark 2008), so we may, in a moment of Dawkins-like fancy, suppose that if technical systems had selves (which they do not), they might see humans are part of *their* extended cognitive systems. In any case, it is now apparent that humans and technical systems are engaged in complex symbiotic relationships, in which each symbiont brings characteristic advantages and limitations to the relationship. The more such symbiosis advances, the more difficult it will be for either symbiont to flourish without the other.

How should we in the humanities analyze and understand this symbiosis? An essential first step is realizing that the human brain has powerful nonconscious cognitive abilities. This realization allows the humanities to see cognition in a new light, not as an ability unique to humans and an attribute virtually synonymous with rationality or higher consciousness, but rather as a capability present in many non-human life-forms and, increasingly, a vast array of intelligent devices. Then the question becomes not whether machines can think, as Alan Turing asked more than a half-century ago, but how networks of nonconscious cognitions between and among the planet's cognizers are transforming the conditions of life, as human complex adaptive systems become increasingly interdependent upon and entwined with intelligent technologies in cognitive assemblages. If contemporary cultures in developed societies are presently undergoing systemic transformations that are profoundly changing planetary cognitive ecologies, as I have argued, then the humanities should and must be centrally involved in analyzing, interpreting, and understanding the implications. Anything less is a disservice to their missions—and to the world. To riff on the quotation with which I began: It all depends on what kind of world you want.

Notes

CHAPTER ONE

1. Research indicates that even before verbal narratives are decoded by higher consciousness, core conscious and perhaps even nonconscious cognition have already initiated physiological reactions appropriate to narrative developments; see, for example, Katrin Riese et al. (2014).

2. For those interested in the Freudian unconscious, it may be regarded as a subset of the "new" unconscious in which some kind of trauma has intervened to disrupt the easy and continuous communication with consciousness. Nevertheless, it manifests to consciousness through dreams and symptoms.

3. Abstract available at https://thesis.library.adelaide.edu.au.

4. In *Plant Theory: Biopower and Vegetable Life*, Jeffrey Nealon argues that plants are the forgotten and even despised subjects in biopolitics and animal rights discourses: "the plant, rather than the animal, functions as that form of life forgotten and abjected within a dominant regime of humanist biopower" (Nealon 2016, location 56).

5. I am indebted for this reference to the anonymous Reader No. 2 for the University of Chicago Press.

6. Cary Wolfe has been (mis)read as alleging that I favor fantasies of disembodiment. In *What Is Posthumanism?* (2009), published eleven years after my *How We Became Posthuman,* he writes: "My sense of posthumanism does not partake of the fantasy of the posthuman described by N. Katherine Hayles, which imagines a triumphant transcendence of embodiment and 'privileges information pattern over material instantiation, so that embodiment in a biological substrate is seen as an accident of history rather than inevitability of life'" (120). In quoting this passage, Wolfe fails to note that although I *describe* this vision, I do so precisely to criticize it (in his Introduction, he is clearer about this distinction [v]). When Wolfe continues the passage quoted above by writing that a fantasy of disembodiment "requires us to attend to that thing called 'the human' with *greater* specificity, *greater* attention to its embodiment, embeddedness, and materiality, and how these in turn shape and are shaped by consciousness, mind, and so on," the effect is to suggest that his position on embodiment opposes my own, when exactly the oppo-

site is the case. Indeed, several people have told me they have (mis)read his remark in precisely this way. To clarify: Wolfe and I agree that human being is grounded in embodiment and embedded in complex social, cultural, and technical milieus. Where we differ is in how the term "posthuman" should be understood. Whereas I identify a spectrum of positions within posthumanism, including ones with which I am in radical disagreement such as fantasies of disembodiment, Wolfe wants to restrict the term "posthuman" to denote only embodied positions, purging it of what he sees as unsavory elements such as fantasies of disembodiment.

7. Hence the argument that Catherine Malabou makes for plasticity over flexibility in the context of human neurology is too narrow to express adequately how flexibility occurs in biological and technical media. For her purposes, of course, plasticity is preferred because flexibility is one of the hallmarks that neoliberal business practices insist workforces must have to remain competitive in global marketplaces, a tactic often used to gloss over job insecurity and the pernicious effects of outsourcing jobs and capital; see Catherine Malabou, *What Should We Do with Our Brain?,* trans. Sebastian Rand (Bronx, N.Y., 2008).

8. Mick Smith, *Against Ecological Sovereignty: Ethics, Biopolitics, and Saving the Natural World* (Minneapolis, 2011), 10.

9. Similar in this regard to Lorenzo Magnani's *Morality in a Technological World: Knowledge as Duty,* Verbeek's approach is less oriented to the kind of "templates of moral doing" (2007, 103) that Magnani employs and more open to the unexpected uses and consequences that technologies may initiate in specific contexts in which humans engage them.

10. See Bruno Latour (2002), "Morality and Technology."

CHAPTER TWO

1. I had a brief introduction to mindfulness techniques on March 20, 2015, while attending "The Total Archive: Dreams of Universal Knowledge from the Encyclopaedia to Big Data" conference at Cambridge University, and this was the phrase used by Matthew Drage, who led us in the mindfulness exercise.

CHAPTER THREE

1. By "material" I mean matter, energy, and information, not only matter in the narrow sense.

2. Whether quantum mechanics can or should be applied to macroscopic objects is a moot point. In the "Science Fiction, Science Fact" class, we discussed the quantum effects of a speeding tennis ball. Mark supplied calculations showing that the de Broglie wavelength (a measure of the ball's wave properties) is on the order of 10^{-32} meters. By comparison, the radius of an atom is 10^{-10} meters, and of a proton, 10^{-15} meters. Not only is 10^{-32} meters a completely immeasurable quantity, but as Mark explained in an e-mail (July 24, 2014), it "does not even make physical sense since measuring

distance to that precision is not possible" because of the Heisenberg Uncertainty Principle.

3. Typical in the renunciation of subjects and signification is this passage from Deleuze and Guattari in *A Thousand Plateaus*: "We are no more familiar with scientificity than we are with ideology; all we know are assemblages. And the only assemblages are machinic assemblages of desire and collective enunciation. No significance, no subjectification: writing to the nth power (all individuated enunciation remains trapped within the dominant significations, all signifying desire is associated with dominated subjects)" (Deleuze and Guattari 1987, 32).

4. Antonio Damasio, for example, posits the proto-self (in my terms, nonconscious cognition) as crucial for consciousness (2000, 174), yet also argues that "there is no self without awareness of and engagement with others" (2000, 194).

5. In this respect, my approach to technical objects shares similarities with media archaeology as defined, for example, by Wolfgang Ernst in *Memory and the Digital Archive* (Minneapolis: University of Minnesota Press, 2012).

CHAPTER FOUR

1. I am indebted to anonymous Reader No. 1 at the University of Chicago Press for this insight.

2. Whether this makes sense physiologically is a moot question. Wisely, McCarthy refrains from specifying the exact nature of the narrator's injury, opting instead to allow readers to infer it from the narrator's actions.

3. Watts says in "Notes and References" (378) that he has read Thomas Metzinger (2004), and the language here clearly reflects Metzinger's ideas.

4. Here we may discern echoes of the controversy over plant intelligence discussed in chapter 1. Since the aliens are like plants in the sense that they have minimal encephalization, their capacities are also vastly underestimated by the brain-directed humans.

5. Note the similarity here to Max Velmans's (1995) description of blindsight. Watts cites Velmans (2003) in "Notes and References."

6. In this sense, the aliens in *Blindsight* can be seen as a sustained exploration of the thesis that Bernard Stiegler (1998) has advanced about the coevolution of humans and technology, arguing that human cognitive capacities are completely entangled with technology from the earliest days of the species.

7. NASA announced the existence of Kepler 452b, in the Cygnus constellation. See Kerry Grens (July 27, 2015).

CHAPTER FIVE

1. A colleague who teaches at UCLA and lives in the San Fernando Valley told me that the dedicated bus lanes had transformed his relation to the university. Whereas previously he was compelled to take the infamous 405 freeway (one of the nation's busiest) in to work, he could now relax on a bus, freed from the tribulations of unpredictable and horrendously snarled traffic.

2. It is worth noting that although he told me the video was available at his website, it has been taken down, and the only mention of the MeMachine is on a Dutch website commemorating an award he received. He may have reasoned that the project was too distracting and was taking attention away from his other scientific work.

3. To their credit, many humanists are doing exactly this. I think of Barbara Stafford, trudging across the University of Chicago campus in the bitter Chicago winter to attend seminars on neuroscience in preparation for her book *Echo Objects: The Cognitive Work of Images* (2008); Deborah Jenson at Duke University, forming a working group involving neuroscientists and humanists; Barbara Herrnstein Smith, creating a Center for Interdisciplinary Studies in Science and Cultural Theory; and, of course, Bruno Latour and Steve Woolgar in their groundbreaking *Laboratory Life: The Construction of Scientific Facts* (1979), as well as many others too numerous to mention.

4. A more reliable estimate is 10 percent. Using a database compiled by New America, a nonpartisan group studying drone warfare, Peter Berger and Jennifer Rowland report this figure (2015, 12–41, especially 18).

5. Some nomenclature clarification is necessary. The technical term for drone is UAV: unmanned aerial vehicle, requiring two pilots, one to guide the craft and the other to monitor sensory data. If more than one aircraft is involved, it becomes UAVS, unmanned aerial vehicle *system*; if intended for combat, UCAVS, unmanned *combat* aerial vehicle system. Autonomous drones are also called UAVs, unmanned *autonomous* vehicles, "aerial" understood from context. To avoid confusion, I will call these UAAVs, unmanned autonomous aerial vehicles, and UAAVS, unmanned autonomous aerial multivehicle systems or swarms.

CHAPTER SIX

1. Yaroufakis adopts the phrase "exorbitant privilege" from Valéry Giscard d'Estaing, President de Gaulle's finance minister from 1959–62, to describe what Yaroufakis characterizes as "the United States' unique privilege to print money at will without any globalized institutionalized constraints" (Yaroufakis, 93).

2. Relevant here is a recent video from the Islamic State purportedly announcing the reissuing of the gold dinar, presented as a "real" currency in contrast to the banknotes of the United States, which the video calls fake or groundless currency. Clearly this propaganda video has its own agenda, but the events it relates about the United States first abandoning the gold standard in the 1930s for domestic currency, and then in 1971 abandoning it for international trade, are factually correct (https://www.youtube.com/watch?v=BG7YXKE4x3w).

3. Automated trading refers to any electronic algorithm that executes buy and sell orders. These may include arbitrage programs that take a day or more to execute a large order. HFT trading is a subset of automated trading; HFT operates in milliseconds and typically results in holdings of very short duration (a few seconds at most) before the product is sold or bought again.

CHAPTER SEVEN

1. Lauren Berlant (2008) agrees with this date, but Ramón Saldívar (2013) places the text's era "well before the heyday of the Civil Rights movement of the 1960s, probably even well before the heroic struggles of the 1940s and '50s" (9). Contra Saldívar, evidence for the later date includes the mayor's sensitivity to the African American population, the hiring of black inspectors in the Department of Elevator Inspectors, and finally, the naming of the Fanny Briggs Memorial Building, because "the mayor is shrewd, and understands that this city is not a Southern city" (12).

2. Some descriptions of Turing's theoretical model have the head moving, others the tape moving. The choice is arbitrary, as long as head and tape move relative to one another. I use the movable head version here to make the analogy with the elevator easier to see. Turing himself imagined that a "computer" (in his time, this meant a person who performs calculations) would move the head/tape according to the program's instructions.

3. Turing himself does not use the word "halting" in his paper but rather translates "Entscheidungsproblem" as "decision problem." It was renamed the Halting Problem by later commentators, a name indicating more precisely the nature of the problem.

4. Procedures for limited classes of algorithms have been devised that will predict whether they will halt; Turing's proof applies to the possibility of a general procedure, applicable to all possible programs that can run on the Turing machine.

CHAPTER EIGHT

1. In mathematics, an inflection point on a curve is the place where the sign of the curvature changes direction, for example, going from concave to convex or vice versa. I use it here as a metaphor for the strategic point at which small differences can have large-scale systemic effects, dramatically altering how a system proceeds in its temporal unfolding.

Works Cited

Appadurai, Arjun. 2016. *Banking on Words: The Failure of Language in the Age of Derivative Finance*. Chicago: University of Chicago Press.

Arkin, Ronald C. 2009. *Governing Lethal Behavior in Autonomous Robots*. Boca Raton, FL: CRC Press.

———. 2010. "The Case for Ethical Autonomy in Unmanned Systems." *Journal of Military Ethics* 9 (4): 332–41.

AR Lab. 2013. "MeMachine: Bio-Technology, Privacy and Transparency." August 13. http://www.arlab.nl/project/memachine-bio-technology-privacy -and-transparency.

Arnuk, Sal, and Joseph Saluzzi. 2012. *Broken Markets: How High Frequency Trading and Predatory Practices on Wall Street Are Destroying Investor Confidence and Your Portfolio*. Upper Saddle River, NJ: FT Press.

Ash, James. 2016. *The Interface Envelope: Gaming, Technology, Power*. London: Bloomsbury Academic.

ATSAC. n.d.: Automated Traffic Surveillance and Control. http://trafficinfo .lacity.org/about-atsac.php. Accessed July 7, 2015.

Auletta, Gennaro. 2011. *Cognitive Biology: Dealing with Information from Bacteria to Minds*. London: Oxford University Press.

Ayache, Elie. 2010. *The Blank Swan: The End of Probability*. Hoboken, NJ: John Wiley.

Baker, R. Scott. 2009. *Neuropath*. New York: Tor Books.

Barad, Karen. 2007. *Meeting the Universe Halfway: Quantum Physics and the Entanglement of Matter and Meaning*. Durham, NC: Duke University Press.

Barsalou, Lawrence W. 2008. "Grounded Cognition." *Annual Review of Psychology* 59: 617–45.

Baucom, Ian. 2005. *Specters of the Atlantic: Finance Capital, Slavery, and the Philosophy of History*. Durham, NC: Duke University Press.

Baudrillard, Jean. 1995. *Simulacra and Simulation*. Ann Arbor: University of Michigan Press.

Beer, Gillian. 1983. *Darwin's Plots: Evolutionary Narrative in Darwin, George Eliot, and Nineteenth-Century Fiction*. Cambridge: Cambridge University Press.

Benjamin, Medea. 2013. *Drone Warfare: Killing by Remote Control*. London: Verso.

Bennett, Jane. 2010. *Vibrant Matter: A Political Ecology of Things*. Durham, NC: Duke University Press.

Bentham, Jeremy. [1780] 1823. *An Introduction to the Principles of Morals and Legislation*. http://www.earlymoderntexts.com/assets/pdfs/bentham1780.pdf.

Berger, Peter L., and Jennifer Rowland. 2015. "Decade of the Drone: Analyzing CIA Drone Attacks, Casualties, and Policy." In *Drone Wars: Transforming Conflict, Law, and Policy*, edited by Peter L. Berger and Daniel Rothenberg, 11–41. Cambridge: Cambridge University Press.

Berlant, Lauren. 2008. "Intuitionists: History and the Affective Event." *American Literary History* 20 (4): 845–60.

Berry, David M. 2011. *The Philosophy of Software: Code and Mediation in the Digital Age* (Kindle Locations 2531–32). London: Palgrave. Kindle edition.

———. 2015. *Critical Theory and the Digital*. London: Bloomsbury Academic.

Bickle, John. 2003. "Empirical Evidence for a Narrative Concept of Self." In *Narrative and Consciousness: Literature, Psychology, and the Brain*, edited by Gary D. Fireman, Ted E. McVay, and Owen J. Flanagan, 195–208. Oxford: Oxford University Press.

Black, Fisher, and Myron Scholes. 1973. "The Pricing of Options and Corporate Liabilities." *Journal of Political Economy* 81 (3): 637–54.

Bogost, Ian. 2012. *Alien Phenomenology, or What It's Like to Be a Thing*. Minneapolis: University of Minnesota Press.

Borges, Jorge Luis. 1994. "Pierre Menard, Author of the *Quixote*." In *Ficciones*, 45–56. Translated by Anthony Kerrigan. New York: Grove Press.

Braidotti, Rosi. 2006. "The Ethics of Becoming Imperceptible." In *Deleuze and Philosophy*, edited by Constantin V. Boundas, 133–59. Edinburgh: Edinburgh University Press.

———. 2013. *The Posthuman*. Malden, MA: Polity Press.

Brenner, Eric D., Rainer Stahlberg, Stefano Mancuso, Jorge Vivanco, František Baluška, and Elizabeth Van Volkenburgh. 2006. "Plant Neurobiology: An Integrated View of Plant Signaling." *Trends in Plant Science* 11 (8): 413–19.

Brenner, Robert. 2006. *The Economies of Global Turbulence: The Advanced Capitalist Economies from Long Boom to Long Turndown, 1945–2005*. Brooklyn, NY: Verso Books.

Bryan, Dick, and Michael Rafferty. 2006. *Capitalism with Derivatives: A Political Economy of Financial Derivatives, Capital and Class*. London: Palgrave Macmillan.

Buchanan, Mark. 2011. "Flash-Crash Story Looks More Like a Fairy Tale." *BloombergView*, May 7. http://www.bloomberg.com/news/2012-05-07/.

Budish, Eric, Peter Cramton, and John Shim. 2015. "The High-Frequency Trading Arms Race: Frequent Batch Auctions as a Market Design Hypothesis." *Quarterly Journal of Economics* 130 (4): 1547–1621. http://faculty.chicagobooth.edu/eric.budish/research/HFT-FrequentBatchAuctions.pdf. Accessed April 7, 2015.

Buffet, Warren. 2002. "Warren Buffet on Derivatives." (Edited excerpts from the Berkshire Hathaway annual report for 2002.) Montgomery Investment

Technology, Inc. http://www.fintools.com/docs/Warren%20Buffet%20on%20Derivatives.pdf.

Burn, Stephen J. 2015. "Neuroscience and Modern Fiction." Special issue of *Modern Fiction Studies* 61 (2): 209–25.

Burroughs, William. 1959. *Naked Lunch*. Paris: Olympia Press.

Callon, Michel. 1998. *Laws of the Market*. Hoboken, NJ: Wiley-Blackwell.

Calude, C. S., and Gregory Chaitin. 1999. "Mathematics: Randomness Everywhere," *Nature* 400 (July 22): 319–20.

Chaitin, Gregory J. 1999. *The Undecidable*. Heidelberg: Springer.

———. 2001. *Exploring Randomness*. Heidelberg: Springer.

———. 2006. *MetaMath! The Quest for Omega*. New York: Vintage.

Chamayou, Grégoire. 2015. *A Theory of the Drone*. Translated by Janet Lloyd. New York: New Press.

Chamovitz, Daniel. 2013. *What a Plant Knows*. New York: Scientific American/Farrar Straus Giroux.

Choudhury, Tanzeem, and Alex Pentland. 2004. "The *Sociometer*: A Wearable Device for Understanding Human Networks." MIT Media Lab. http://alumni.media.mit.edu/~tanzeem/TR-554.pdf.

Clark, Andy. 1989. *Microcognition: Philosophy, Cognitive Science, and Parallel Distributed Processing*. 2nd ed. Cambridge, MA: MIT Press.

———. 2008. *Supersizing the Mind: Embodiment, Action, and Cognitive Extension*. London: Oxford University Press.

Coeckelbergh, Mark. 2011. "Is Ethics of Robotics about Robots? Philosophy of Robotics beyond Realism and Individualism." *Law, Innovation and Technology* 3 (2): 241–50.

Corballis, Michael C. 2015. *The Wandering Mind: What the Brain Does When You're Not Looking*. Chicago: University of Chicago Press.

Damasio, Antonio. 2000. *The Feeling of What Happens: Body and Emotion in the Making of Consciousness*. New York: Mariner Books.

———. 2005. *Descartes' Error: Emotion, Reason, and the Human Brain*. New York: Penguin. Originally published 1995.

———. 2012. *Self Comes to Mind: Constructing the Conscious Brain*. New York: Vintage Books.

Dehaene, Stanislas. 2009. "Conscious and Nonconscious Processes: Distinct Forms of Evidence Accumulation." *Séminaire Poincaré* 12:89–114. http://www.bourbaphy.fr/dehaene.pdf.

———. 2014. *Consciousness and the Brain: Deciphering How the Brain Codes Our Thoughts*. New York: Penguin.

Dehaene, Stanislas, Claire Sergent, and Jean-Pierre Changeux. 2003. "A Neuronal Network Model Linking Subjective Reports and Objective Physiological Data during Conscious Perception." *Proceedings of the National Academy of Sciences USA* 100:8520–25.

Deleuze, Gilles, and Félix Guattari. 1987. *A Thousand Plateaus: Capitalism and Schizophrenia*. Translated by Brian Massumi. Minneapolis: University of Minnesota Press.

Dennett, Daniel C. 1992. *Consciousness Explained*. New York: Back Bay Books.

Dresp-Langley, Birgitta. 2012. "Why the Brain Knows More Than We Do:

Non-Conscious Representations and Their Role in the Construction of Conscious Experience." *Brain Science* 2 (1): 1–21.

Dreyfus, Hubert. 1972. *What Computers Can't Do*. Cambridge, MA: MIT Press.

———. 1992. *What Computers Still Can't Do*. Cambridge, MA: MIT Press.

———. 2013. "The Myth of the Pervasiveness of the Mental." In *Mind, Reason, and Being-in-the-World: The McDowell-Dreyfus Debate*, edited by Joseph K. Schear, 15–40. London: Routledge.

Dupuy, Jean-Pierre. 2009. *On the Origins of Cognitive Science: The Mechanization of Mind*. Cambridge, MA: MIT Press.

Eagleman, David. 2012. *Incognito: The Secret Lives of the Brain*. New York: Random House.

Edelman, Gerald M. 1987. *Neural Darwinism: The Theory of Neuronal Group Selection*. New York: Basic Books.

Edelman, Gerald M., and Giulio Tononi. 2000. *A Universe of Consciousness: How Matter Becomes Imagination*. New York: Basic Books.

Ekman, Ulrik. 2015. "Design as Topology: U-City." In *Media Art and the Urban Environment: Engendering Public Engagement with Urban Ecology*, edited by Francis T. Marchese, 177–203. New York: Springer.

Encyclopedia Britannica. n.d. "Cognition." www.brittannica.com/topic /cognition-thought-process.

Epstein, R. S. 2014. "Mobile Medical Applications: Old Wine in New Bottles?" *Clinical Pharmacology and Therapeutics* 95 (5): 476–78.

Ernst, Wolfgang. 2012. *Memory and the Digital Archive*. Minneapolis: University of Minnesota Press.

Fazi, M. Beatrice. 2015. "The Aesthetics of Contingent Computation: Abstraction, Experience, and Indeterminacy." PhD diss., Goldsmiths University of London.

Fish, Stanley. 2012. "Mind Your P's and B's: The Digital Humanities and Interpretation." *Opinionator* (blog). *New York Times*, January 23. http://opinionator.blogs.nytimes.com/2012/01/23/mind-your-ps-and-bs-the-digital-humanities-and-interpretation/?_r=0.

Fisher, Mark. 2009. *Capitalist Realism*. New York: Zero Books.

Flanagan, Owen. 1993. *Consciousness Reconsidered*. Cambridge, MA: MIT Press.

Freeman, Walter J., and Rafael Núñez. 1999. "Editors' Introduction." In *Reclaiming Cognition: The Primacy of Action, Intention and Emotion*, edited by Rafael Núñez and Walter J. Freeman, ix–xix. New York: Imprint Academic.

Friedman, Norman. 2010. *Unmanned Combat Air Systems: A New Kind of Carrier Aviation*. Annapolis, MD: Naval Institute Press.

Fuller, Matthew. 2007. *Media Ecologies: Materialist Energies in Art and Technology*. Cambridge, MA: MIT Press.

Galloway, Alexander R., and Eugene Thacker. 2007. *The Exploit: A Theory of Networks*. Minneapolis: University of Minnesota Press.

Gardner, Martin. 1970. "The Fantastic Combinations of John Conway's New Solitaire Game 'Life.'" *Scientific American* 223 (October): 120–23.

Gitelman, Lisa. 2014. *Paper Knowledge: Toward a Media History of Documents*. Durham, NC: Duke University Press.

Gladwell, Malcolm. 2005. *Blink: The Power of Thinking without Thinking.* New York: Little, Brown.

Gleick, James. 2012. *The Information: A History, a Theory, a Flood.* New York: Vintage.

Goldsmith, Kenneth. 2008. "Conceptual Poetics." *Harriet* (a poetry blog). Poetry Foundation. http:/www.poetryfoundation.org/harriet/2008/06 /conceptual-poetics-kenneth-goldsmith/?woo.

Goodwin, Brian C. 1977. "Cognitive Biology." *Communication and Cognition* 10 (2): 87–91.

Gould, Stephen Jay. 2007. *Punctuated Equilibrium.* Cambridge, MA: Belknap Press.

Grens, Kerry. 2015. "Most Earth-Like Planet Found." *Scientist,* July 27. http://www.the-scientist.com/?articles.view/articleNo/43605/title/Most-Earth-like-Planet-Found/.

Grosz, Elizabeth. 2002. "A Politics of Imperceptibility: A Response to 'Antiracism, Multiculturalism and the Ethics of Identification.'" *Philosophy of Social Criticism* 28:463–72.

———. 2011. *Becoming Undone: Darwinian Reflections on Life, Politics, and Art.* Durham NC: Duke University Press.

Guillory, John. 2008. "How Scholars Read." *ADE Bulletin* 146 (Fall): 8–17.

Haddon, Mark. 2004. *The Curious Incident of the Dog in the Night-Time.* New York: Vintage.

Han, Jian, Chang-hong Wang, and Guo-xing Yi. 2013. "Cooperative Control of UAV Based on Multi-Agent System." In *Proceedings of the 2013 IEEE 8th Conference on Industrial Electronics and Applications (ICIEA): 19–21 June 2013, Melbourne, Australia.* Piscataway, NJ: IEEE. http://ieeexplore.ieee.org/xpl /login.jsp?tp=&arnumber=6566347&url=http%3A%2F%2Fieeexplore.ieee .org%2Fxpls%2Fabs_all.jsp%3Farnumber%3D6566347.

———. 2014. "UAV Robust Strategy Control Based on MAS." *Abstract and Applied Analysis,* vol. 2014. Article ID 796859.

Hansen, Mark B. N. 2015. *Feed-Forward: On the Future of Twenty-First Century Media.* Chicago: University of Chicago Press.

Harman, Graham. 2011. *The Quadruple Object.* Abington, Oxon: Zero Books.

Hassin, Ran R., James S. Uleman, and John A. Bargh, eds. 2005. *The New Unconscious.* Oxford: Oxford University Press.

Hayles, N. Katherine. 2010. "Cybernetics." *Critical Terms for Media Studies,* edited by W. J. T. Mitchell and Mark B. N. Hansen, 145–56. Chicago: University of Chicago Press.

———. 2012. *How We Think: Digital Media and Contemporary Technogenesis.* Chicago: University of Chicago Press.

———. 2014a. "Cognition Everywhere: The Rise of the Cognitive Nonconscious and the Costs of Consciousness." *New Literary History* 45 (2): 199–220.

———. 2014b. "Speculative Aesthetics and Object Oriented Inquiry (OOI)." *Speculations: A Journal of Speculative Realism* 5 (May). http://www .speculations-journal.org/?page_id=5.

———. 2016. "The Cognitive Nonconscious: Enlarging the Mind of the Humanities." *Critical Inquiry* 42 (4): 783–808.

Heimfarth, Tales, and João Paulo de Araujo. 2014. "Using Unmanned Aerial Vehicle to Connect Disjoint Segments of Wireless Sensor Network." *IEEE 28th International Conference on Advanced Information Networking and Applications (AINA), 2014: 13–16 May 2014, University of Victoria, Victoria, Canada; Proceedings*, edited by Leonard Barolli, 907–14. Piscataway, NJ: IEEE.

Ho, Karen. 2009. *Liquidated: An Ethnography of Wall Street*. Durham, NC: Duke University Press.

Horowitz, Eli, Matthew Derby, and Kevin Moffett. 2014. *The Silent History*. New York: Farrar, Straus and Giroux.

Human Rights Watch. 2012. "Losing Humanity: The Case against Killer Robots." Cambridge, MA: International Human Rights Clinic, Harvard Law School.

Hutchins, Edwin. 1996. *Cognition in the Wild*. Cambridge, MA: MIT Press.

Iliadis, Andrew. 2013. "Informational Ontology: The Meaning of Gilbert Simondon's Concept of Individuation." *Communication +1*, vol. 2, article 5. http://scholarworks.umass.edu/cpo/vol2/iss1/5/.

James, William. 1997. *The Meaning of Truth*. Amherst, NY: Prometheus Books. Originally published 1909.

Jameson, Fredric. 2010. "Realism and Utopia in *The Wire*." *Criticism* 52 (3–4): 359–72.

Johnson, Neil, Guannan Zhao, Eric Hunsader, Jing Meng, Amith Ravindar, Spencer Carran, and Brian Tivnan. 2012. "Financial Black Swans Driven by Ultrafast Machine Ecology." Physics and Society Working Paper, Cornell University. www.arXiv.1202.1448.

Johnson, Neil, Guannan Zhou, Eric Hunsader, Hong Qi, Nicholas Johnson, Jing Meng, and Brian Tivran. 2013. "Abrupt Rise of New Machine Ecology beyond Human Response Time." *Scientific Reports: Nature Publishing Group*, Article 2627: 1–11. http://www.nature.com/articles/srep02627.

Johnston, John. 2008. "*The Intuitionist* and *Pattern Recognition*: A Response to Lauren Berlant." *American Literary History* 20 (4): 861–69.

Jonze, Spike. 2012. *Her*. Burbank, CA: Warner Home Video. DVD.

Kandel, Eric R., and James H. Schwartz. 2012. *Principles of Neural Science*. 5th ed. New York: McGraw-Hill Education.

Kelly, James Floyd. 2012. "Book Review and Author Interview: *Kill Decision* by Daniel Suarez." *Wired Magazine*, July 7. http://archive.wired.com/geekdad /2012/07/daniel-suarez-kill-decision/.

Kelly, Kevin. 2010. *What Technology Wants*. New York: Penguin.

Kouider, Sid, and Stanislas Dehaene. 2007. "Levels of Processing during Nonconscious Perception: A Critical Review of Visual Masking." *Philosophical Transactions of the Royal Society* B 362:857–75.

Kováč, Ladislav. 2000. "Fundamental Principles of Cognitive Biology." *Evolution and Cognition* 6 (1): 51–69. http://dai.fmph.uniba.sk/courses/CSCTR /materials/CSCTR_03sem_Kovac_2000.pdf.

———. 2007. "Information and Knowledge in Biology: Time for Reappraisal." *Plant Signalling and Behavior* 2 (March–April): 65–73.

Koza, John R. 1992. *Genetic Programming: On the Programming of Computers by Means of Natural Selection*. Cambridge, MA: Bradford Books, MIT Press.

Krishnan, Armin. 2009. *Killer Robots: Legality and Ethicality of Autonomous Weapons*. Farnham, UK: Ashgate Publishing Limited.

Lakoff, George, and Mark Johnson. 2003. *Metaphors We Live By*. 2nd ed. Chicago: University of Chicago Press.

Lange, Ann-Christina. 2015. "Crowding of Adaptive Strategies: High-Frequency Trading and Swarm Theory." Presentation at the "Thinking with Algorithms" Conference, the University of Durham, UK, February 27.

Langton, Christopher, ed. 1995. *Artificial Life*. Cambridge, MA: MIT Press.

Latour, Bruno. 1992. "Where Are the Missing Masses? Sociology of a Few Mundane Artifacts." In *Shaping Technology-Building Society. Studies in Sociotechnical Change*, edited by Wiebe Bijeker and John Law, 225–59. Cambridge, MA: MIT Press.

———. 1999. *Pandora's Hope: Essays on the Reality of Science Studies*. Cambridge, MA: Harvard University Press.

———. 2002. "Morality and Technology: The End of the Means." Translated by Couze Venn. *Theory, Culture and Society*, 19 (5–6): 247–60.

———. 2007. *Reassembling the Social: An Introduction to Actor Network Theory*. London: Oxford University Press.

Latour, Bruno, and Steve Woolgar. 1979. *Laboratory Life: The Construction of Scientific Facts*. Princeton, NJ: Princeton University Press.

Lem, Stanislaw. 2014. *Summa Technologiae*. Translated by Joanna Zylinska. Minneapolis: University of Minnesota Press.

Lenoir, Timothy, and Eric Giannella. 2011. "Technology Platforms and Layers of Patent Data." In *Unmaking Intellectual Property: Creative Production in Legal and Cultural Perspectives*, edited by Mario Biagioli, Peter Jaszi, and Martha Woodmansee. Chicago: University of Chicago Press.

Lethem, Jonathan. 2000. *Motherless Brooklyn*. New York: Vintage.

Levinas, Emmanuel. 1998. *Otherwise Than Being: On Beyond Essence*. Pittsburgh: Duquesne University Press.

Levy, Steven. 2014. "Siri's Inventors Are Building a Radical New AI That Does Anything You Ask." *Wired Magazine*, August 12. http://www.wired.com/2014/08/viv/.

Lewicki, Pawel, Thomas Hill, and Maria Czyzewska. 1992. "Nonconscious Acquisition of Information." *American Psychology* 47 (6): 796–801.

Lewis, Michael. 2014. *Flash Boys: A Wall Street Revolt*. New York: W. W. Norton.

Libet, Benjamin, and Stephen M. Kosslyn. 2005. *Mind Time: The Temporal Factor in Consciousness*. Cambridge, MA: Harvard University Press.

Ling, Philip. 2010, "Redefining Firmware." *New Electronics*, January 11. http://www.newelectronics.co.uk/electronics-technology/redefining-firmware/21841/.

López-Maury, L., S. Marguerat, and J. Bähler. 2008. "Tuning Gene Expression to Changing Environments: From Rapid Responses to Evolutionary Adaptation." *Nature Reviews Genetics* 9 (8): 583–93.

Lowenstein, Roger. 2001. *When Genius Failed: The Rise and Fall of Long-Term Capital Management*. New York: Harper Collins.

Lynch, Deidre Shauna. 1998. *The Economy of Character: Novels, Market Culture, and the Business of Inner Meaning*. Chicago: University of Chicago Press.

Lyon, Pamela C., and Jonathan P. Opie. 2007. "Prolegomena for a Cognitive Biology." Presented at the Proceedings of the 2007 Meeting of International Society for the History, Philosophy and Social Studies of Biology, University of Exeter. Abstract at https://digital.library.adelaide.edu.au/dspace /handle/2440/46578. Accessed June 10, 2015.

MacKenzie, Donald. 2003. "Long-Term Capital Management and the Sociology of Arbitrage." *Economy and Society* 32 (2): 349–80.

———. 2005. "How a Superportfolio Emerges: Long-Term Capital Management and the Sociology of Arbitrage." In *The Sociology of Financial Markets*, edited by Karin Knorr Cetina and Alex Preda, 62–83. New York and London: Oxford University Press.

———. 2008. *An Engine, Not a Camera: How Financial Models Shape Markets*. Cambridge, MA: MIT Press.

———. 2011. "How to Make Money in Microseconds." *London Review of Books* 33 (10): 16–18.

Magnani, Lorenzo. 2007. *Morality in a Technological World: Knowledge as Duty.* Cambridge: Cambridge University Press.

Malabou, Catherine. 2008. *What Should We Do with Our Brain?* Translated by Sebastian Rand. Bronx, NY: Fordham University Press.

Marcus, Ben. 2012. *The Flame Alphabet*. New York: Vintage.

Marcus, Sharon. 2013 "Description and Critique." Paper presented at "Interpretation and Its Rivals" Conference, University of Virginia, September.

Margulis, Lynn, and Dorian Sagan. 1986. *Microcosmos: Four Billion Years of Evolution from Our Microbial Ancestors*. New York: Summit Books.

Marshall, Kate. 2014. "The View from Above." Paper presented at the 129th Modern Language Association Convention, Chicago, IL, January 9–12.

Maturana, Humberto R., and Francisco J. Varela. 1980. *Autopoiesis and Cognition: The Realization of the Living*. Dordrecht: D. Reidel Publishing.

Mauer, Bill. 2002. "Repressed Futures: Financial Derivatives' Theological Unconscious." *Economy and Society* 31 (1): 15–36.

Maxwell, James Clerk. 1871. *Theory of Heat*. London: Longmans.

McCarthy, Tom. 2007. *Remainder*. New York: Vintage.

McDowell, John. 1996. *Mind and World*. With a new introduction by the author. Cambridge, MA: Harvard University Press.

———. 2013. "The Myth of the Mind as Detached." In *Mind, Reason, and Being-in-the-World: The McDowell-Dreyfus Debate*, edited by Joseph K. Schear, 41–58. London: Routledge.

McEwan, Ian. 1998. *Enduring Love: A Novel*. New York: Anchor.

McLemee, Scott. 2013. "Crunching Literature." Review of *Macroanalysis: Digital Methodsand Literary History*, by Matthew L. Jockers. *Inside Higher Ed*, May 1. https://www.insidehighered.com/views/2013/05/01/review-matthew-l -jockers-macroanalysis-digital-methods-literary-history.

Meillassoux, Quentin. 2010. *After Finitude: An Essay on the Necessity of Contingency*. London: Bloomsbury Academic.

Metzinger, Thomas. 2004. *Being No One: The Self-Model Theory of Subjectivity*. Cambridge, MA: MIT Press.

Mitchell, Tom. n.d. "NELL: Never-Ending Language Learning." Read the Web:

Research Project at Carnegie Mellon University. rtw.ml.cmu.edu/rtw. Accessed October 17, 2015.

Moretti, Franco. 2013. "Network Theory, Plot Analysis." In *Distant Reading*, 211–40. New York: Verso.

Nealon, Jeffrey. 2016. *Plant Theory: Biopower and Vegetable Life*. Stanford, CA: Stanford University Press.

Neefjes, J., and R. van der Kant. 2014. "Stuck in Traffic: An Emerging Theme in Diseases of the Nervous System." *Trends in Neuroscience* 37 (2): 66–76.

Nelson, Katherine. 2003. "Narrative and the Emergence of a Consciousness of Self." In *Narrative and Consciousness: Literature, Psychology and the Brain*, edited by Gary D. Fireman, Ted E. McVay, and Owen J. Flanagan, 17–36. Oxford: Oxford University Press.

Neocleous, Mark. 2003. "The Political Economy of the Dead: Marx's Vampires." *History of Political Thought* 24, no. 4 (Winter).

Nicolelis, Miguel. 2012. *Beyond Boundaries: The New Neuroscience of Connecting Brains with Machines—and How It Will Change Our Lives*. Reprint ed. London: St. Martins Griffin.

Núñez, Rafael, and Walter Freeman, eds. 1999. *Reclaiming Cognition: The Primacy of Action, Intention and Emotion*. Thorverton, UK: Imprint Academic.

Otis, Laura. 2001. *Networking: Communicating with Bodies and Machines in the Nineteenth Century*. Ann Arbor: University of Michigan Press.

Parikka, Jussi. 2010. *Insect Media: An Archaeology of Animals and Technology*. Minneapolis: University of Minnesota Press.

Parisi, Luciana. 2004. *Abstract Sex: Philosophy, Bio-Technology and the Mutations of Desire*. London: Continuum.

———. 2015. "Critical Computation: Digital Philosophy and GAI." Presentation at the "Thinking with Algorithms" Conference, University of Durham, UK, February 27. Forthcoming in *Theory, Culture and Society*.

Parisi, Luciana, and Steve Goodman. 2011. "Mnemonic Control." In *Beyond Biopolitics: Essays on the Governance of Life and Death*, edited by Patricia Ticento Clough and Craig Willse, 163–76. Durham, NC: Duke University Press.

Patterson, Scott. 2010. *The Quants: How a New Breed of Math Whizzes Conquered Wall Street and Nearly Destroyed It*. New York: Random House.

———. 2012. *Dark Pools: High-Speed Traders, A.I. Bandits, and the Threat to the Global Financial System*. New York: Crown Business.

Paycheck. 2003. Director John Woo.

PBS. 2010. "The Warning." Director Michael Kirk. *Frontline*. Available at http://www.pbs.org/wgbh/pages/frontline/warning/view/.

———. 2013. "Rise of the Drones." *Nova*, January 23. http://www.pbs.org/wgbh/nova/military/rise-of-the-drones.html.

Pentland, Alex. 2008. *Honest Signals: How They Shape Our World*. Cambridge, MA: MIT Press.

Perez, Edgar. 2011. *The Speed Traders: An Insider's Look at the New High-Frequency Phenomenon That Is Transforming the Investing World*. New York: McGraw Hill.

Perloff, Marjorie. 2012. *Unoriginal Genius: Poetry by Other Means in the New Century*. Chicago: University of Chicago Press.

Pinker, Steven. 2007. *The Language Instinct: How the Mind Creates Language.* New York: Harper.

Pollan, Michael. 2013. "The Intelligent Plant." *New Yorker*, December 23. http://www.newyorker.com/magazine/2013/12/23/the-intelligent-plant.

Poovey, Mary. 2008. *Genres of the Credit Economy: Mediating Value in Eighteenth- and Nineteenth-Century Britain.* Chicago: University of Chicago Press.

Power, Matthew. 2013. "Confessions of a Drone Warrior." *GQ*, October 23. http://www.gq.com/news-politics/big-issues/201311/drone-uav-pilot -assassination.

Powers, Richard. 2007. *The Echo Maker.* New York: Picador.

Ramachandran, V. S. 2011. *The Tell-Tale Brain: A Neuroscientist's Quest for What Makes Us Human.* New York: W. W. Norton.

Regan, Tom. 2004. *The Case for Animal Rights.* Berkeley: University of California Press.

Riese, Katrin, Mareike Bayer, Gerhard Lauer, and Annekathrin Schact. 2014. "In the Eye of the Recipient." *Scientific Study of Literature* 4 (2): 211–31. http://www.ingentaconnect.com/content/jbp/ssol/2014/00000004/00000002 /art00006.

Rosch, Eleanor. 1999. "Reclaiming Concepts." In *Reclaiming Cognition: The Primacy of Action, Intention and Emotion*, edited by Rafael Núñez and Walter J. Freeman, 61–78. Thorverton, U.K: Imprint Academic.

Rosen, Robert. 1991. *Life Itself: A Comprehensive Inquiry into the Nature, Origin, and Fabrication of Life.* New York: Columbia University Press.

Rotman, Brian. 1987. *Signifying Nothing: The Semiotics of Zero.* Stanford, CA: Stanford University Press.

Rowe, E. 2002. "The Los Angeles Automated Traffic Surveillance and Control (ATSAC) System." *Vehicular Technology, IEEE Transactions* 40 (1): 16–20.

Sacks, Oliver. 1998. *The Man Who Mistook His Wife for a Hat: And Other Clinical Tales.* New York: Touchstone.

Saldívar, Ramón. 2013. "The Second Elevation of the Novel: Race, Form, and the Postrace Aesthetic in Contemporary Narrative." *Narrative* 21 (1):1–18.

Schear, Joseph K., ed. 2013. *Mind, Reason, and Being-in-the-World: The McDowell-Dreyfus Debate.* London: Routledge.

Scott, David. 2014. *Gilbert Simondon's Psychic and Collective Individuation: A Critical Introduction and Guide.* Edinburgh: Edinburgh University Press.

Searle, John. 1980. "Minds, Brains and Programs." *Behavioral and Brain Sciences* 3 (3): 417–57.

Serres, Michel, with Bruno Latour. 1995. *Conversations on Science, Culture and Time.* Ann Arbor: University of Michigan Press.

Shannon, Claude E. 1993. "Prediction and Entropy of Printed English." In *Claude E. Shannon: Collected Papers*, edited by N. Sloane and A. Wyner, 194–208. New York: Wiley-IEEE Press.

Shannon, Claude E., and Warren Weaver. 1948. *The Mathematical Theory of Communication.* New York: American Telephone and Telegraph Co.

Shukin, Nicole. 2009. *Animal Capital: Rendering Life in Biopolitical Times.* Minneapolis: University of Minnesota Press.

Simondon, Gilbert. 1989. *L'individuation psychique et collective: À la lumière*

des notions de forme, information, potentiel et métastabilité. Paris: Editions
  Aubier.
Simons, Daniel J., and Christopher S. Chabis. 1999. "Gorillas in Our Midst."
  Perception 28:1059–74.
———. 2011. *The Invisible Gorilla: How Our Intuitions Deceive Us.* New York:
  Harmony Books.
Singer, Peter W. 2010. "The Ethics of Killer Applications: Why Is It So Hard to
  Talk about Morality When It Comes to New Military Technology?" *Journal
  of Military Ethics* 9 (4): 299–312.
Smith, Mick. 2011. *Against Ecological Sovereignty: Ethics, Biopolitics, and Saving
  the Natural World.* Minneapolis: University of Minnesota Press.
Stafford, Barbara. 2008. *Echo Objects: The Cognitive Work of Images.* Chicago:
  University of Chicago Press.
Stewart, Garrett. 1990. *Reading Voices: Literature and the Phonotext.* Berkeley:
  University of California Press.
Stiegler, Bernard. 1998. *Technics and Time,* vol. 1: *The Fault of Epimetheus.*
  Translated by Richard Beardsworth and George Collins. Stanford, CA:
  Stanford University Press.
———. 2008. *Technics and Time,* vol. 2: *Disorientation.* Translated by Stephen
  Barker. Stanford, CA: Stanford University Press.
———. 2010a. *For a New Critique of Political Economy.* Cambridge: Polity
  Press.
———. 2010b. *Taking Care of Youth and the Generations.* Stanford, CA: Stanford
  University Press.
Stockdale Center for Ethical Leadership, US Naval Academy, McCain Confer-
  ence. 2010. "Executive Summary and Command Brief." *Journal of Military
  Ethics,* 9 (4): 424–31.
Stone, Christopher D. 2010. *Should Trees Have Standing? Law, Morality, and the
  Environment.* 3rd ed. London: Oxford University Press. Originally published
  1972.
Strang, Veronica. 2014. "Fluid Consistencies: Material Relationality in Human
  Engagements with Water." *Archeological Dialogues* 21 (2): 133–50.
Suarez, Daniel. 2013. *Kill Decision.* New York: Signet.
Taleb, Nassim Nicholas. 2010. *The Black Swan: The Impact of the Highly Improb-
  able.* 2nd ed. New York: Random House.
Tamietto, Marco, and Beatrice de Gelder. 2010. "Neural Bases of the Non-
  conscious Perception of Emotional Signals." *Nature Reviews* 11 (October):
  697–709.
Terranova, Tiziana. 2006. "The Concept of Information." *Theory, Culture and
  Society* 23:286.
Thrift, Nigel. 2004. "Remembering the Technological Unconscious by Fore-
  grounding Knowledges of Position." *Environment and Planning D. Society
  and Space* 22:175–90.
———. 2007. *Non-Representational Theory: Space, Politics, Affect.* London: Routledge.
Tompkins, Peter, and Christopher Bird. 1973. *The Secret Life of Plants.* New
  York: Harper and Row.

Trewavas, A. 2005. "Aspects of Plant Intelligence." *Annals of Botany* (London) 92:1–20.

Tucker, Patrick. 2014. "Inside the Navy's Secret Swarm Robot Experiment." *Defense One* (October 5). http://www.defenseone.com/technology/2014/10 /inside-navys-secret-swarm-robot-experiment/95813/. Accessed July 7, 2015.

Turing, Alan. 1936–37. "On Computable Numbers, with an Application to the Entscheidungsproblem." *Proceedings of the London Mathematical Society*, ser. 2, 42:230–85. http://www.turingarchive.org/browse.php/B/12.

University of Massachusetts Medical School. n.d. "Mindfulness-Based Stress Reduction (MBSR)." http:www.umassmed.edu/cfm/stress/index.aspx.

Van der Helm, Frans. 2014. "Design/Embodiment." Panel discussion at the Critical and Clinical Cartographies Conference, Delft University of Technology, November 13–14. http://www.bk.tudelft.nl/fileadmin/Faculteit/BK /Actueel/Agenda/Agendapunten_2014/doc/Critical_Clinical_Cartographies _Conference_brochure.pdf. Accessed July 7, 2015.

Varela, Francisco J., and Paul Bourgine, eds. 1992. *Toward a Practice of Autonomous Systems*. Cambridge, MA: MIT Press.

Varela, Francisco J., Evan Thompson, and Eleanor Rosch. 1992. *The Embodied Mind: Cognitive Science and Human Experience*. Cambridge, MA: MIT Press.

Vasko, Timothy. 2013. "Solemn Geographies of Human Limits: Drones and the Neocolonial Administration of Life and Death." *Affinities: A Journal of Radical Theory, Culture, and Action* 6 (1): 83–107.

Velmans, Max. 1995. "The Relation of Consciousness to the Material World." In *Explaining Consciousness: The Hard Problem*, edited by J. Shear. Cambridge, MA: MIT Press. Available at http://www.meta-religion.com /Philosophy/Articles/Consciousness/relation_of_consciousness.htm.

———. 2003. "Preconscious Free Will." *Journal of Consciousness Studies* 10 (2): 42–61.

Verbeek, Peter-Paul. 2011. *Moralizing Technology: Understanding and Designing the Morality of Things*. Cambridge: Cambridge University Press.

Von Neumann, John. 1966. *Theory of Self-Reproducing Automata*. Edited and completed by Arthur W. Banks. Urbana: University of Illinois Press.

Wall, Cynthia Sundberg. 2014. *The Prose of Things: Transformations of Description in the Eighteenth Century*. Chicago: University of Chicago Press.

Watts, Peter. 2006. *Blindsight*. New York: Tor Books.

Weiskrantz, Lawrence, E. K. Warrington, M. D. Sanders, and J. Marshall. 1974. "Visual Capacity in the Hemianopic Field Following a Restricted Occipital Ablation." *Brain* 97:709–28.

Whitehead, Alfred North. 1978. *Process and Reality: An Essay on Cosmology*. Edited by David R. Griffin and Donald W. Sherburne. New York: Free Press/ Macmillan.

Whitehead, Colson. 1999. *The Intuitionist*. New York: Anchor Books.

———. 2009. "Year of Living Postracially." Op-Ed. *New York Times*, November 2.

Wiener, Norbert. 1950. *The Human Use of Human Beings: Cybernetics and Society*. New York: Houghton Mifflin.

Williams, Raymond. 1977. *Marxism and Literature*. Oxford: Oxford University Press.

————. 2003. *Television: Technology and Cultural Form*. London: Routledge.

Wilson, E. O. 2014. *The Meaning of Human Existence*. New York: W. W. Norton.

Wolfe, Cary. 2009. *What Is Posthumanism?* Minneapolis: University of Minnesota Press.

Yaroufakis, Yanis. 2013. *The Global Minotaur: America, Europe, and the Future of the Global Economy*. London: Zed Books.

Zenko, Micah. 2013. *Reforming U.S. Drone Strike Policy* (Council Special Report). Washington, DC: Council on Foreign Relations Press.

Index